T0201581

MODELING AND SIMULATION IN THE MEDICAL AND HEALTH SCIENCES

MODELING AND SIMULATION IN THE MEDICAL AND HEALTH SCIENCES

Edited by

John A. Sokolowski, Ph.D.
Catherine M. Banks, Ph.D.

The Virginia Modeling, Analysis and Simulation Center
Old Dominion University
Suffolk, Virginia

WILEY

A JOHN WILEY & SONS, INC., PUBLICATION

Published by John Wiley & Sons, Inc., Hoboken, New Jersey.
Published simultaneously in Canada.

For general information on our other products and services or for technical support, please contact our Customer Care Department within the United States at (800) 762-2974, outside the United States at (317) 572-3993 or fax (317) 572-4002.

Wiley also publishes its books in a variety of electronic formats. Some content that appears in print may not be available in electronic formats. For more information about Wiley products, visit our web site at www.wiley.com.

Library of Congress Cataloging-in-Publication Data Is Available

Modeling and simulation in the medical and health sciences / edited by John A. Sokolowski and Catherine M. Banks

  Includes bibliographical references and index.

  ISBN 978-0-470-76947-8

Printed in Singapore

oBook ISBN: 9781118003206
ePDF ISBN: 9781118003183
ePub ISBN: 9781118003190

10 9 8 7 6 5 4 3 2 1

My wife, Marsha, and my daughters, Amy and Whitney,
for all your love and support
—John A. Sokolowski

My husband, James, the love of my life
—Catherine M. Banks

CONTENTS

10   Future of Modeling and Simulation in the Medical and Health Sciences   175

Richard M. Satava

Appendix   Modeling Human Behavior, Modeling Human Systems: Addressing the Skepticism, Responding to the Reservations   195

CONTRIBUTORS

JOHN A. ADAM, Department of Mathematics and Statistics, Old Dominion University, Norfolk, Virginia

CATHERINE M. BANKS, Virginia Modeling, Analysis and Simulation Center, Old Dominion University, Suffolk, Virginia

C. DONALD COMBS, Eastern Virginia Medical School, Norfolk, Virginia

MOHAMMED FERDJALLAH, Virginia Modeling, Analysis and Simulation Center, Old Dominion University, Suffolk, Virginia

ELIZABETH A. JACOB, Thayer School of Engineering, Dartmouth College, Hanover, New Hampshire

LINDSAY B. KATONA, Hitchcock Medical Center, Dartmouth College, Hanover, New Hampshire

GYUTAE KIM, Department of Electrical and Computer Engineering, Old Dominion University, Norfolk, Virginia

KEUM JOO KIM, Thayer School of Engineering, Dartmouth College, Hanover, New Hampshire

RICHARD LEE, Department of Robotics Surgery, Englewood Hospital and Medical Center, Englewood, New Jersey

DEQING LI, Thayer School of Engineering, Dartmouth College, Hanover, New Hampshire

PAUL E. PHRAMPUS, University of Pittsburgh Medical School, Pittsburgh, Pennsylvania

STACIE I. RINGLEB, Department of Mechanical Engineering, Old Dominion University, Norfolk, Virginia

JOSEPH M. ROSEN, Hitchcock Medical Center, Dartmouth College, Hanover, New Hampshire

EUGENE SANTOS, JR., Thayer School of Engineering, Dartmouth College, Hanover, New Hampshire

RICHARD M. SATAVA,   Department of Surgery, University of Washington, Seattle, Washington

JOHN A. SOKOLOWSKI,   Virginia Modeling, Analysis and Simulation Center, Old Dominion University, Suffolk, Virginia

FEI YU,   Thayer School of Engineering, Dartmouth College, Hanover, New Hampshire

FOREWORD

Advances in the medical and health sciences are partly the result of the variety of modeling and simulation tools available to professionals in these fields. The technology now facilitates accurate representations of the body that serve as baseline models for assessing and prescribing courses of actions for disturbances to the human system. To fully engage this capability, medical professionals – users – need to be trained to use this technology. They must also participate in the development of this technology to ensure that these tools meet their standards.

Modeling and Simulation in the Medical and Health Sciences introduces this modeling and simulation application domain from both the engineer's and medical professional's perspective to provide just that – a better understanding of simulation development for medical applications. Throughout the chapters an underlying theme is presented: the call for greater collaboration between simulation developers and users.

The discussion purposely introduces a broad review of simulation in the medical and health sciences. First, it presents the necessary introductory material in the field of modeling and simulation, then it progresses to modeling itself such as mathematical representations of human elements and the employment of several modeling techniques designed to replicate humans and human systems. The concluding chapters address a variety of uses of simulation in training and patient care.

Medical professionals are keenly aware of the need to broaden medical simulation applications with a view to expedited and effective training of healthcare providers.

Significant also to medical simulation applications is its ability to facilitate better informed decision-making during critical junctures in patient care – as the text calls it, *facilitating a seamless mode of information transmission.* To develop effective training tools and application tools, effort is needed to bring together multidisciplinary expertise. *Modeling and Simulation in the Medical and Health Sciences* is the right place to start.

ARYEH SHANDER, MD, FCCM, FCCP

Englewood Hospital and Medical Center
Mount Sinai School of Medicine
November 2010

PREFACE

With the technology boom of the 1990s came varied approaches to modeling, varying degrees of simulation, and sophisticated methods of visual representation—essentially an informal introduction to what would eventually comprise the discipline of modeling and simulation (M&S). Students in the engineering and computer science disciplines were the first beneficiaries of these technological advancements, and it wasn't long before that technology, coupled with an expanding body of knowledge on modeling and simulation, fast-tracked across the disciplines. This placed M&S at the forefront of a multidisciplinary effort to integrate quantitative and qualitative research methods and diverse modeling paradigms. What is more, M&S now possesses a variety of modeling tools that can represent many aspects of life, including life itself. The medical and health sciences are proof of that.

M&S is providing practitioners in these fields with the capability to better understand some of the fundamental aspects of health care, such as human behavior, human systems, medical treatment, and disease proliferation. Whether live, virtual, or constructive modes of M&S are used, nurse educators and physicians are trained in a variety of applied areas through simulations developed from mathematical, physical, computer, and human models. Thus, it can be said that *medical and health sciences M&S is an evolutionary, interdisciplinary process of model development and simulation design requiring the expertise of developers (M&S experts) and users (medical and health care trainers and practitioners) to facilitate a seamless mode of information transmission.* This book provides a venue to do just that, as it is designed to educate future members of the medical M&S community toward developing and perfecting a seamless mode of information transmission in the health care domain.

In our view, the phrase *seamless mode of information transmission* has meaning on two levels. First and foremost is the sole objective of medical and health sciences M&S: to create environments whereby precise information *transfers directly to* or is *discovered by* the health care provider. Academicians and commercial developers of medical and health sciences M&S products are working on this crucial goal by creating the ideal virtual operating room, the perfect prosthesis, and the best diagnostic imaging apparatus for *users*. Essentially, for medical and health sciences M&S to have any significant impact in health care, a seamless mode of information transmission via models and simulations must take place. The second meaning, and one that serves as the impetus for this book, is that M&S *developers* and *users* must share expertise, requirements, and criticisms while recognizing limitations and expectations regarding model development and simulation design. Modeling is not easy, and the human body is one of the most complex systems for modelers to attempt. Similarly, medical trainers and practitioners recognize the dynamism of the body; therefore, they cannot always provide discrete, static portraits of the anatomy or quantitatively convey degrees of pain. Understanding the parameters and the tasks that both audiences address in their work is critical if the goal of a seamless mode of information transmission is to be accomplished. This publication facilitates that understanding by providing an interdisciplinary study for future members of the health care M&S community toward a greater capacity for collaboration with M&S developers.

This book is intended for engineering graduate students focusing their research on the development of medical and health sciences M&S and for medical and health care students who will be engaging M&S as practitioners or trainers. The content is an orientation to the theory and applications of M&S in the medical and health sciences. The book can also serve as a valuable resource for medical and health sciences majors who desire a technical understanding of modeling and simulation as they might function as future consultants to M&S developers. All readers of this book will no doubt benefit from the incisive analysis of the key concepts, body of knowledge, and M&S applications in medical and health sciences provided by the scholars and expert practitioners who will have contributed to the book.

Note, in particular, that the discussion has been set within the reasonable bounds of a graduate course in medical and health sciences modeling and simulation by introducing key concepts, citing the body of knowledge, elaborating on theoretical modeling underpinnings, and referencing M&S applications in research and education. Thus, we make no claims that it is an exhaustive study of model development, simulation design, and applications of M&S in the medical and health sciences. Plainly omitted from this discussion are a number of subsets: *adaptive serious games* (training and rehabilitation), *E-health*, *clinical lab maintenance*, *insurance systems modeling*, *medical informatics* (hospital information systems, computer-based patient records), *clinical engineering* (such as risk factors, safety, and management of medical equipment), *quality improvement and team building*, *hospital design*, *open standards for medical M&S*, and *ethical issues* associated with use of medical technology.

Students studying disciplines in the sciences (e.g., computer science, mathematics, biology) and engineering (e.g., mechanical, bioengineering, electrical, systems, M&S) are grounded in the fundamentals of M&S but lack an understanding of the medical applications, such as the human body as a system, disease proliferation, and even gait analysis. These students need to become versed in understanding the human system, in understanding what takes place as training in the medical and health sciences, and in advancing the three modes of M&S: live, virtual, and constructive. We focus on that necessary M&S education via a multidisciplinary approach, with chapter contributions from faculty across the disciplines as well as medical experts possessing both Ph.D. and medical degrees.

With the M&S student as *developer* in mind, we approach the topic of medical and health sciences M&S in three phases as a way to introduce approach, theory, and application in a methodical manner. Thus, we start by introducing the fundamental modes of M&S, progress to explaining the theoretical origins of the model, and conclude by highlighting M&S treatment in research and in education and utilization.

The book is divided into three parts, beginning with a general discussion of M&S as a discipline. In Part One, *"Fundamentals of Medical and Health Sciences M&S,"* we ground the student in common terminology. We also relate this terminology to concepts employed in the medical and health care area to help bridge the gap between developers and practitioners. Three distinct modes of M&S are described: live, constructive, and virtual. The **live** approach explains the concept of using real (live) people employing real equipment for training purposes. The **constructive** mode is a means of *engaging* medical M&S. In constructive simulation, simulated people and simulated equipment are developed to augment real-world conditions for training or experimentation purposes. The **virtual** mode is perhaps the most fascinating, as virtual operating rooms and synthetic training environments are being produced for practitioners and educators at breakneck speed. In this mode, real people employ simulated equipment to improve physical skills and decision-making ability.

The development of any model takes its form from a theoretical perspective. In Part Two, *"Modeling for the Medical and Health Sciences,"* we discuss both computational and physical models. **Computational models** exist in either a purely mathematical form such as a series of equations or in an algorithmic form implemented in a digital computer. **Physical models** can be manikins that contain representations of human anatomy for the purpose of practicing surgical or diagnostic procedures. Computational and physical models directly support the modes of M&S described above. Computational models are most closely associated with the constructive mode; physical models are used from a virtual mode perspective. These linkages are explored in the introductory chapters of the book.

Part Three, *"Modeling and Simulation Applications,"* covers two general areas: (1) medical and health sciences M&S research, such as the use of humans as models and human systems modeling, and (2) medical and health sciences M&S education, such as in the areas of robotics, training, and patient care. There is a wide range of medical and health sciences M&S research. We look first at humans as models and then at the challenge and manner of modeling the human system. **Humans as models** makes use of real people, who may be asked to do things such as portray or mimic particular disease symptoms to provide an interactive diagnosis training experience for medical and health sciences students and professionals. This type of modeling is included in the live mode. **Human systems modeling** introduces analytical and computational methods to model and simulate medical principles as a way of understanding how the organ systems control the functions of the body. This type of modeling integrates expertise from numerous disciplines, including biology, mathematics, and computer science.

An overview of medical and health sciences M&S education as a general application area is useful in assessing the current tools available and the need to refine or expand that toolbox. The use of **robotics** for invasive surgical procedures is making inroads in many hospitals. Surgeons trained to use these tools are observing significant decreases in patient recovery. Robotics is one among numerous M&S applications being used for **training.** All aspects of care taking can be taught in a fully immersive virtual operating room fitted with a simulated patient and both real and simulated equipment. The system is designed to provide training in judgment and decision making for members of surgical teams using both real and virtual team members. M&S technology is also being used to augment training with standardized patients: persons who are used to portray patients realistically, to teach and assess communication and other clinical skills. Stethoscopes allow the learner to hear abnormal heart and lung sounds when placed on a normal, healthy standardized patient. Another medical and health sciences M&S application area, much more explicit, is **patient care**, which is the culmination of medical and health sciences M&S research, development, and training.

Although the illustrations are not printed in color, some chapters have figures that are described using color. The color representations of these figures may be downloaded from the following site: ftp://ftp.wiley.com/public/sci_tech_med/modeling_simulation.

JOHN A. SOKOLOWSKI
CATHERINE M. BANKS

PART ONE
Fundamentals of Medical and Health Sciences Modeling and Simulation

1 Introduction to Modeling and Simulation in the Medical and Health Sciences

CATHERINE M. BANKS

INTRODUCTION

Technological advancements have paved the way for new approaches to modeling, simulation, and visualization. Modeling now encompasses high degrees of complexity and holistic methods of data representation. Various levels of simulation capability allow for improved outputs and analysis of discrete and continuous events, and state-of-the-art visualization allows for graphics that can represent details within a single shaft of hair [1]. These technological developments were first exploited among the engineering and computer science disciplines; however, the expanding body of knowledge and user-friendly applications of modeling and simulation (M&S) have resulted in applications across the disciplines. As such, M&S is at the forefront of multidisciplinary collaboration that integrates quantitative and qualitative research methods and diverse modeling paradigms. Significantly, these modeling tools are capable of representing many aspects of life, including life itself. Case in point: the use of M&S in the medical and health sciences (MHSs).

Practitioners in the MHSs are engaging M&S to explore and understand some fundamental aspects of health care, such as human behavior, human systems, medical treatment, and disease proliferation. The training tools available to people in these fields include the three primary modes of M&S: live, virtual, and constructive. These modes facilitate the development of mathematical, physical, computer, and human models. Thus, it can be said that *medical and health sciences is an evolutionary, interdisciplinary process of model development and simulation design requiring the expertise of developers (M&S*

Modeling and Simulation in the Medical and Health Sciences, First Edition. Edited by John A. Sokolowski and Catherine M. Banks.
© 2011 John Wiley & Sons, Inc. Published 2011 by John Wiley & Sons, Inc.

experts) *and users* (*medical and health care trainers and practitioners*) *to facilitate a seamless mode of information transmission.* The information discussed in this book is designed to educate future members of the MHS M&S community toward developing and perfecting a seamless mode of information transmission in the health care domain via M&S.

Consider this seamless mode of information transmission as having two interrelated meanings. The first is a focus on basic M&S as it pertains to the MHSs, in which the objective is to create environments whereby precise information *transfers directly to* or is *discovered by* the health care provider. The second meaning, and one that serves as the impetus for this book, is that M&S *developers* and *users* must share expertise, requirements, and criticisms while recognizing limitations and expectations regarding model development and simulation design. Any expert modeler will freely admit that modeling is not easy: The more complex the system or entity to be represented or characterized, the more difficult the task of modeling it. Added to that is the difficulty of modeling the organic, dynamic nature of the human body. Similarly, medical trainers and practitioners recognize the dynamism of the body; therefore, they cannot always provide discrete, static portraits of the anatomy or quantitatively convey degrees of pain. Therefore, both developers and users must appreciate the parameters and the tasks that each one encounters to best facilitate a seamless mode of information transmission. In this chapter we introduce the current challenges of developing and engaging M&S in the MHSs. We present the role of M&S as two complementary activities in health care studies: the development of tools and the training and use of those tools. An introductory overview of these concepts is a good place to start.

MODELING AND SIMULATION IN THE MEDICAL AND HEALTH SCIENCES

The M&S body of knowledge is expanding as interest in the discipline and application of models and simulations increases. The academic programs in which the core curriculum of M&S is taught are found in the engineering and computer science departments. These disciplines dominate the body of literature, which is grounded in mathematics, engineering, and computer science. As M&S applications and user friendliness increases, so will student (and user) interest. For many students M&S serves as a way to explore hypotheses and serves as a training tool. This has also been the case with students studying medicine and health sciences; the long history of medical modeling is proof of that.

There is, however, a growing concern that a lack of understanding exists on the part of the MHS student when using M&S solely as a training tool. This deficiency creates a void in understanding the theoretical underpinnings of the M&S. Conversely, engineering students as M&S developers must appreciate the fact that acceptance of medical applications of M&S depends on such issues as performance, robustness, and accuracy—attributes that require medical expertise and/or input at the development stage. A cursory review of the MHS M&S literature sheds light on the fact that a gap exists in the scholarship that disconnects developer and user.

As a whole, the MHS M&S body of knowledge is comprised of books, journals, and conference proceedings that span both the development and application sides of the domain. Probably the most complete MHS M&S subarea in the literature is the technical–developer community responsible for *visualization and imaging*. Academic training for visualization and imaging is generally found in the electrical and computer engineering and/or computer science disciplines, both of which have been perfecting the development of visualization and imaging technology. In fact, there is a long tradition of scientists and engineers who illustrate their work with graphics that include anatomical illustrations and computer images to provide representations to store three-dimensional geometry and efficient algorithms that render these representations. Other developer-side contributions to the body of literature include subareas such as *biomedical* and *devices and systems: technology and informatics*, which speak to computational intelligence and medical simulation as well as to developing next-generation tools for medical education and patient care.

Biomedical engineering is the application of engineering concepts and techniques to problems in medicine and health care. This is a relatively new domain with typical applications in prosthetics, medical instruments, diagnostic software, and imaging equipment. Computational intelligence techniques consist of computing algorithms and learning machines, including neural networks, fuzzy logic, and genetic algorithms. One such study designed for graduate-level students is the 2008 Begg et al. text discussing state-of-the-art applications of computational intelligence in cardiology, electromyography, electrocephalography, movement science, and biomechanics [2]. Numerous biomedical handbooks are available. A notable text is *Medical Devices and Systems* edited by Bronzino [3], which introduces the term **clinical engineer**. These engineers are closely aligned with **biomedical engineers**, whose primary focus areas include the development of biocompatible prostheses; various diagnostic and therapeutic medical devices such as clinical equipment to microimplants; common imaging equipment such as MRIs and EEGs; biotechnologies such as regenerative tissue growth; and pharmaceutical drugs and biopharmaceuticals. Clinical engineers also apply electrical, mechanical, chemical, optical, and engineering principles to understand, modify, or control biologic systems; and they assist in diagnosis and treatment.

Also found on the user–educator side of the body of knowledge is a significant series entitled *Medicine Meets Virtual Reality* [4–6]. These edited volumes of short papers present different *approaches to* and *uses of* simulation. As a whole the series is committed to knowledge sharing and building bridges via breakthrough applications in simulation, visualization, robotics, and informatics as well as experiences between physicians in all specialties, scientists in various disciplines, educators, and even commercial entities that serve as retailers of this technology. This *bridge building* is significant and necessary. Additionally, the series includes cognitive and behavioral assessments derived from simulation trials used for examining a variety of scenarios, ranging from enhancing dental treatment processes to examining schizophrenia.

Among the numerous essays in the Westwood et al. volumes is one of special interest to M&S educators, as it explains the need to develop body of knowledge repositories and commonly agreed upon definitions for the medical M&S vocabulary. In "Visualizing the Medical Modeling and Simulation Database: Trends in the Research Literature," the authors, Combs and Walia, present a structured categorization of the literature (choosing to bin it into eight categories) as well as general terminology that can serve as a baseline for a common lexicon, such as *procedural simulation* and *telemedicine* [4].

All students of MHS M&S need to stay current with this expanding body of literature to understand the basic theoretical underpinnings of M&S technology and the new tools available to practitioners. These tools include **engineered devices** such as the cochlear implant, the defibrillator, and the pacemaker, as well as novel applications stemming from the field of **biomechatronics**, which merges humans with machines. **Robotics** is also providing innovative approaches to the human–machine interface as well as in clinical procedures. Computer-based M&S has led the development of **training simulations** where medical practitioners can hone their skills and expand their experience through the use of haptic devices, digital models, and imaging capability.

The body of literature draws attention to the divide or gap that exists between the technical engineer who develops M&S tools and the practitioner who applies the methods. Therein lies the challenge: to balance of both worlds. Students who master the challenge will have accomplished a necessary, meaningful, and useful contribution to these disciplines. A good place to start that mastery is a basic understanding of the terminology and concepts relevant to the study of M&S in the MHSs.

Definition of Basic Terms and Concepts

Throughout this book the reader will be introduced to a number of terms and concepts in the MHS domain. Appropriate for this chapter is a brief introduction to some of these terms and concepts, beginning with the fundamental modes of M&S and the general nature of the model itself. (See Chapter 2 for a detailed review of this information.)

The discipline of M&S and the use of M&S applications is grounded primarily on analysis, experimentation, and training. **Analysis** refers to the investigation of a model's behavior. **Experimentation** occurs when the behavior of the model changes under conditions that exceed the design boundaries of the model. **Training** is the development of knowledge, skills, and abilities obtained as one operates the system represented by the model. There are three **modes of M&S**—live, virtual, and constructive—and they are the same no matter what discipline makes use of M&S. Any discussion of these modes should originate from the discipline of the M&S perspective. This facilitates the establishment of a common terminology. It also relates this terminology to concepts found in the MHS domains to help bridge the gap between developers and MHS practitioners. First, there is the **live mode** approach, the concept of using real (live) people employing real equipment for training purposes. Next, the **virtual mode**, which is perhaps the most fascinating, as virtual operating rooms and synthetic training environments are being produced for practitioners and educators at breakneck speed. In this mode, real people are employing simulated equipment to improve physical skills and decision-making ability. Finally, there is the **constructive mode**, used as a means of *engaging* MHS M&S. In constructive simulation, simulated people and simulated equipment are developed to augment real-world conditions for training or experimentation purposes.

All modeling originates from a theoretical perspective, and it evolves from a conceptual model. The nature of the model can be computational or physical. **Computational models** exist in a purely mathematical form such as a series of equations, or in an algorithmic form implemented in a digital computer. **Physical models** can be manikins that contain representations of human anatomy used for the purpose of practicing surgical or diagnostic procedures. Computational and physical models directly support the three modes of M&S. However, computational models are most closely associated with the constructive mode, whereas physical models are commonly engaged in a virtual mode. (These modeling natures are discussed in detail in Chapters 3 and 4.)

It is important that developers of MHS M&S understand the modes and the nature (or origin) of a model. This information is necessary in determining what type of model would best serve for MHS training or as a practitioner's tool. Training and tools are the two primary categories in which many MHS applications of M&S can be found.

Modeling and Simulation Applications

Generally speaking, the application of models and simulations are found in two broad areas. The first area is research, which encompasses such things as humans as models, human systems modeling, and disease modeling. **Humans as models** makes use of real people so that they can portray or mimic particular disease symptoms to provide an interactive diagnosis training experience for MHS students and professionals. Naturally, this type of modeling is included in the live mode. (This type of modeling is discussed in detail in Chapter 5). Conversely, **human systems modeling** introduces analytical and computational methods to model and simulate medical principles as a way of understanding how the organ systems control functions of the body. This type of modeling is interdisciplinary, as it makes use of expertise from numerous disciplines, such as biology, mathematics, and computer science. (This topic is reviewed in detail in Chapter 6.)

The second area is usage and education, such as in training and patient care. The use of mechanical means to facilitate less invasive surgical procedures—robotics—is discussed in Chapter 7. Much of the M&S education in an MHS relies on current tools that are used primarily as training applications: for example, a fully immersive **virtual operating room** fitted with a simulated patient and both real and simulated equipment. The operating room is designed to provide training in judgment and decision making for members of surgical teams using both real and virtual team members. M&S technology is also engaged to augment training with **standardized patients** (i.e., people who realistically portray patients—used to teach and assess communication and other clinical skills). **Modified stethoscopes** allow the learner to hear abnormal heart and lung sounds when placed on a normal, healthy, standardized patient. (The topic of training is addressed in Chapter 8).

Patient care deals with using simulation to improve or contribute directly to a patient's overall health and well-being. Simulation makes it possible to test new protocols and to design new products to achieve an improved level of health. (See Chapter 9 for a discussion of patient care.)

RESEARCH AND DEVELOPMENT OF MEDICAL AND HEALTH SCIENCES M&S

As M&S educators, the editors of this book found it interesting that the MHS M&S community, a subset of the M&S community at large, would refer to the *science of simulation* as a separate entity or discipline. This encouraged the editors to conduct an analysis of M&S in the MHSs in an attempt to bring together under one umbrella a shared lexicon. Frankly, until researching the resources for this textbook study, neither editor had heard the term *science of simulation* (and Dr. Sokolowski has a Ph.D. in modeling and simulation!). As such, a book by Kyle and Murray, *Clinical Simulation: Operations, Engineering, and Management*, drew attention [7]. That text addresses simulation as a core element of training in medicine, surgery, clinical care, biomedical engineering, and the medical sciences. Specific to this study was an enlightening essay by Richard M. Satava of the University of Washington.

Satava speaks of the collective support for using simulation as a medical educational tool on the part of the Residency Review Committee of the Accreditation Council on Graduate Medical Education. The council now requires that residency programs have simulation as an integral part of their training programs. The American College of Surgeons has also recognized this transformation and is certifying training centers to ensure the quality of simulation training provided. For medical professionals serving as educators, simulation has become what Satava calls *a training environment with permission to fail*, where students are *taught by errors*. Thus, as part of a broader M&S education program, this book is designed to encourage the M&S developer community to take hold of the *science of simulation* subset and integrate it into the M&S body of research and development. To separate the two is confusing and nonproductive, and it perpetuates a disconnect between developer and user. Only recently have the challenges posed by this dichotomy been realized.

In the 2005 report by the National Academy of Engineering, the call for engineering and health care partnerships rang out loudly. By February 2009, medical simulation legislation had been enacted. Appropriately called *SIMULATION* (Safety In Medicine Utilizing Leading Advanced Simulation Technologies to Improve Outcomes Now), this legislation extends the benefits of advanced medical simulation technology to the civilian health care system and calls for the establishment of simulation technology in medical, nursing, allied health, podiatric, osteopathic, and dental education and training protocols.

M&S has been grappling with accurately characterizing or representing human behavior in models developed from real-world case studies. This is a challenge because human behavior is difficult to evaluate (as it is unpredictable and dynamic) and to quantify. The human body is equally unpredictable and dynamic; thus, it is difficult to model from a developer side. The inherent uncertainty within the model and/or the simulation tools calls to question user confidence. Yet, one must ask: What is the alternative? Simply refuse to model, or attempt to model, human systems or case studies that reflect degrees of uncertainty? No. M&S educators are compelled to press forward with enhancing modeling capability in an effort that best represents the human factors and the human system while fully recognizing the limitations of M&S and the fluidity of the entity being modeled. A multidisciplinary approach to MHS M&S research and development fosters ongoing enhancement and improvement in the applications and tools available for the user community. As the MHS M&S toolbox expands and becomes more sophisticated, a *training environment with permission to fail* will yield both a desired *seamless mode of information transmission* and *proficient medical practitioners*.

MHS Research Centers

M&S is now an established research and development domain that is supported at all levels of government. In July 2007 the U.S. House of Representatives unanimously passed House Resolution 487, declaring M&S a *national critical technology* that can provide unparalleled advancements in American competitiveness, develop new and innovative ways to protect the homeland, and bring high-tech jobs and economic prosperity to our communities. Among the numerous descriptors found in the resolution is one specific to MHS: "acknowledges the significant impacts of M&S on a breadth of fields including defense, space, national disaster response, medicine, transportation, and construction." The same 2009 House and Senate resolutions that introduced *SIMULATION* (HR.855/S.616) focused further attention on the fact that M&S expertise would be needed at all levels: development, assessment (verification and validation), and usability.

From an engineering perspective, there is now, more than ever, a need for partnership between engineers and health care professionals; this is so if engineers as developers and medical professionals as users are to meet the six goals outlined by the Institute of Medicine in Washington, DC and its focus on *21st Century Health-Care*. The report by the institute calls for the health care system to be safe, effective, patient-centered, timely, efficient, and equitable. This is a challenge because the health care system is experiencing overwhelming advances and threatening declines. While the changes in medical technology and practice are advancing and improving training and patient care, shortages of skilled health care professionals are becoming severe.

Throughout the United States there are numerous institutes and centers whose research and development focus on *user* MHS M&S. The Advanced Initiatives in Medical Simulation—**AIMS**—is a coalition of professionals and organizations intent on promoting medical simulation to improve patient safety, reduce medical errors, train, and reduce health care costs. This collaborative body endeavors to engage the MHS M&S community, articulate a community-wide message to foster a uniform community, and secure resources necessary for future research and development. The Center for Integration of Medicine and Innovative Technology—**CIMIT**—is another leader in the MHS M&S community. Its aim is to initiate and accelerate translational medical research in the domain of devices, procedures, and clinical systems engineering. Harvard's Center for Medical Simulation—**CMS**—provides simulation training for health care providers through high-fidelity scenarios. (There are numerous other centers worthy of note, including Computer Aided Surgery, CAS; Image Sciences Institute, ISS; Computer Assisted Radiology and Surgery, CARS; and the Society in Europe of Simulation Applied to Medicine, SESAM.) Conversely, modeling and simulation research and development centers, such as the Virginia Modeling, Analysis, and Simulation Center—**VMASC**—also support various MHS M&S application areas and domains.

At VMASC, faculty among numerous disciplines (engineering, mathematics, health sciences, sciences) work in four areas of research and development: (1) training, (2) treatment, (3) disease modeling, and (4) management of health care systems. In the area of training, VMASC researchers developed a fully immersive virtual operating room outfitted with a simulated patient and both real and simulated instruments. In the domain of treatment, M&S is engaged for rehabilitation, to improve the diagnosis and treatment of orthopedic injuries and disorders, and to optimize physical performance. Mathematical models and computer simulations covering a variety of diseases are used for disease modeling. In the subset management of health care systems, researchers are using M&S to understand the effects of bioterrorism on the health care system in conjunction with a mass casualty model.

Meeting the growing demands of MHS M&S effectively and efficiently requires that the education of the M&S developer include explicit input from the user community—the MHS practitioner. A solid foundation for that type of education has its roots in a multidisciplinary approach: developers and users in the same classroom. As technology and application improve, simulation will no doubt find a permanent home in the MHSs. For many developers this is a grand accomplishment; for many users this is an opportunity to provide highly sophisticated patient care. But for some, these advancements give rise to the ethical question of using simulation, inanimate technology, to represent animate behavior and human systems.

THE QUESTION OF ETHICS IN MEDICAL AND HEALTH SCIENCES M&S

The advances made in computer capability (software, artificial intelligence, and software agents) facilitate the simulation of complex phenomena such as human behavior modeling and the modeling of human systems.[†] Modeling human systems encompasses the representation of the human body. This requires great technical skill, for with greater accuracy of human physiology comes higher-fidelity, more realistic simulation. For the most part, medical simulation is becoming an accepted methodology for educating future medical practitioners and for providing ongoing training and assessment for practicing professionals.

Juxtaposed with these technical advances is the ethical concern of representing human physiology via simulation. The concern emanates from academic disciplines that rely heavily on soft (or fuzzy) and evolving data such as the social sciences and medical and health sciences. Developers (both those producing the models and those creating the simulations) must address the credibility and ethical concern of modeling dynamic organic systems such as the human body to validate M&S as a MHS practitioners' tool. Central to the ethical discussion are the questions: Should computer models premised on mathematics represent human action (behavior) and human physiology (systems)? Is simulation a valid and verifiable means of characterizing humans? Discussing these questions proffers a means to bridge developer and user understanding of M&S. It also serves to communicate both the constraints and the potential of M&S.

The multidimensional capability of M&S credits it with being an enabling technology. As such, subfields of MHS make wide use of simulation, such as training with manikins, to using haptic devices, to imaging devices and virtual operating rooms.[‡] There is a plethora of applications with promise of even more as medical simulation is used in multiple corners: research, evidence-based outcomes, medical education, and performance assessment, to name a few. In the MHS community medical simulation takes on the term **surrogacy**, which refers to the three modes of simulation in a clinical setting as human (a.k.a live), virtual, and mechanical (a.k.a constructive) [8]. Recall that human (live) simulation uses a trained role-player to act the part of a patient with a specific medical condition. This has been problematic in that a major limitation of human simulation is the inability of students to perform invasive procedures and other therapeutic interventions that could be harmful to the role-player. Virtual simulation employs the latest advances

---

[†]For a detailed history of medical simulation, see "Modeling and Simulation: Real World Examples—Medical" by C. Donald Combs, in *Principles of Modeling and Simulation: A Multidisciplinary Approach* edited by John A. Sokolowski and Catherine M. Banks, Wiley, Hoboken, NJ, 2009.

[‡]Human behavioral modeling allows for the incorporation of socially dependent aspects of behavior that occur when a member of people are together. Human behavior modeling is used in fields of study that include observations of human behavior, be they individual, group, or crowd.

in computer technology and visual interfaces to create acceptably realistic learning experiences.

Game-based medical simulation is proving to be an efficient and effective tool for teaching many clinical lessons without real patients and instructors. In **game-based learning** (GBL), realism is created in hybrid platforms that allow students to explore virtual environments with highly sophisticated mechanical interfaces. Additionally, there is *simulated vision* such as three-dimensional photorealism, which provides a sensation of realistic vision in virtual simulation. To create *simulated touch*, **haptic** devices are used to facilitate the sense of touch—students are able to feel what they are doing as well as to see it. This teaches the student how to gauge pressure application. Mechanical (constructive) simulation allows students to use mock or artificial parts to mimic the experience that would typically be gained from interacting with a real patient's body, organs, or tissues. In **augmented reality**, an application of integrated technology simulation is used to enhance real therapeutic interventions. Those endorsing augmented reality believe that *reality plus simulation produces a better outcome than reality alone.* Still, professionals in the MHS view the implied trade-off between reality and simulation as insufficient proof that one is superior to the other [8]. So what role should simulation play in MHS?

In 2001, a simulation gaming symposium drafted an assessment of the current state of affairs in MHS M&S [9]. Included was the general concurrence that the bridges between medicine and simulation are many and varied. Coupled with that is the fact that medical simulation itself is in overlapping areas of medicine and health care. Practitioners across the large domain of health care for the most part consider medical simulation beneficial. They contend that as the population increases, awareness of health issues improves, society continues to age, and MHS research and development progresses, health care demands will increase exponentially. Thus, the symposium came to what it believed to be a logical conclusion: There is a need to exploit the benefits of simulation in both training and operation.

Pedagogically, many institutions are using M&S training tools specific to MHS. This is due to the fact there is a great need for doctors and health care providers to quickly gain and maintain competence and demonstrate proficiency in the use of these new technologies. Consider the following:

- It is predicted that by 2020 there will be a 20% shortage of nurses.
- Medical residents training or operating within an 80-hour workweek have less opportunity to interact with instructors.
- It is estimated that deaths from medical errors range from 44,000 to 98,000 annually, with 1 million injuries attributed to medical error.
- Various threats (such as bioterrorism) call attention to the speed needed to train first responders (EMS personnel) to react to health crises.

One can speculate as to why these numbers are so high: lack of adequate training, overworked personnel, inadequate tools, and insufficient staffing, among others. A proponent of simulation training, S. B. Issenberg (Center for Research in Medical Education, University of Miami School of Medicine), has suggested a framework for thinking about how to use medical simulators. Issenberg emphasizes the use of repetition, measurement of performance, and feedback as a way to strengthen and to standardize important components of medical education [10]. Health care educational institutions are also aware of the need to provide students with simulated experiences that will enhance educational experience and improve patient safety [11].[†] Still, this type of education and training is not without its critics, who raise questions of validity and ethics.

For example, there are numerous potential applications of simulation for the assessment of clinical competence; however, that use has not been widely supported within the health care community [11]. Concerns about the use of simulation to performance assessment include competing tensions between dual goals: One goal is to achieve high reliability and the other goal is to achieve high validity. The concern is that these goals appear to require mutually exclusive test conditions. Still, experimental studies have demonstrated that moderate levels of reliability and validity can be achieved simultaneously with manikin simulations if test conditions are managed appropriately.

Other critics of simulation use have made the case that simulation not be used in isolation to make a determination about a practitioner's overall performance; rather, it should be incorporated into a broader, multifaceted program [11]. There are also high-stakes performance assessments. These assessments refer to those that lead to an overall judgment about a previously qualified practitioner's professional performance, for the purpose of renewing practice privileges. These high-stakes performance assessments may involve standardized screening exercises for routine recertification, or they can be a tailored exercise. Proponents of this application make the case that it is beneficial to do this in a simulated environment because using real patients is considered unethical, there are not enough patients available for these exercises, and there are not enough cases to provide realistic patient variability [11]. Training with cadavers and laboratory animals is challenging. Cadavers have the correct anatomy; however, they are expensive, sometimes difficult to procure, and suffer tissue degradation. Animals pose a challenge as well: They do not have the same anatomy, can be expensive, and are not reusable. (There is also an ethical question regarding the use of both cadavers and animals in medical training, such as that expressed by PETA, People for

---

[†] Haptic technology serves as the interface within a simulated environment, engaging the user's sense of touch by applying forces, vibrations, and/or motions to the user.

Ethical Treatment of Animals.) For health care educators, simulations allow the practice of safety, prevention, containment, treatment, and procedure in a risk-free setting [7].

Simulation provides a method enveloping both training and feedback in which students can practice a task in lifelike circumstances using models or virtual reality with an opportunity to gather feedback; this, in turn, allows students to practice and review as often as required to attain the desired level of proficiency. Arguing for the integration of simulation in MHS education, McGaghie, et al. [12] refute some of the issues presented by critic Geoffrey Norman, editor of *Advances in Health Sciences Education* [13]. Norman listed seven issues, ranging from the lack of a common platform for sharing, due to the fact that the discipline is so new; the fact that the training outcomes in using real patients and simulated tools are only slightly better; and the fact that the costs of high-fidelity simulations limit which schools can afford them.

Richard M. Satava of the University of Washington has also supported the use of simulation in MHS education. Satava considers simulation an appropriate medical educational tool. His conclusion coincides with the Residency Review Committee of the Accreditation Council on Graduate Medical Education, which has begun requiring residency programs to have simulation as an integral part of their training programs. The American College of Surgeons (ACS) is also making use of the tool and is certifying training centers to ensure the quality of training provided. Satava suggests that the by-product of simulation use in health care education will be standardized curricula [7].[†] As such, simulation-based medical education (SBME) has been expanding its tools and approaches to include task trainers, simulated patients, computer-based training, and full-environment simulators.

That *permission to fail* (as described by Satava) learning environment is also an evidence-based learning environment [7]. This is because simulation allows for anything that reproduces experimental, clinical, or educational data. Still, one must be careful to note the difference between the model and the simulation. Herein lies the ethical and validity question because *a simulation is no better than the model it engages.* Also, the model is arguably the most important part of the simulation, as it constrains the simulation from being a false representative. Thus, clinical simulators need a model to drive them, and that model development is called into question if and when the

[†]For a discussion on simulation use in surgical education, see "Support for Simulation-Based Surgical Education Through American College of Surgeons: Accredited Education Institutes," by A. K. Sachdeva, C. A. Pellegrini, and K. A. Johnson in *World Journal of Surgery* (2008) 32:196–207, DOI 10.1007/s00268-007-9306-x. As of 2008, the American College of Surgeons certification requires three categories of students who are to be taught using simulation.

user is doing the modeling, especially if the user is not a trained modeler. This is a concern, because what one sees in the simulator is what the model produced. Simply, if the physiology (the model) isn't realistic, the simulation isn't realistic.

To address this concern, simulations must be able to pass the **Turing test**, which means that under a variety of circumstances there is no discernible difference between the output of the model in a simulation and the true condition of real patient [14].[†] Thus, a clinical simulator should have an accurate physiological model as its engine. This is necessary because a failed Turing test can teach inaccurate information to students. Also acknowledged is the fact (or engineering constraint) that even the best model is a compromise, as not every detail can be included. *Model compromise* is an art, and those constructing models for clinical simulation should have detailed knowledge of physiology, pharmacology, toxicology, and clinical realism [7]. Unfortunately, this has not always been the case, leading some users to disavow the use of simulation.

The prevalent philosophy in medical ethics emanates from two perspectives: beneficence and nonmaleficence [3]. **Beneficence** speaks to the provision of benefits, whereas **nonmaleficence** centers on the avoidance of doing harm. Traditionally, the ethics of medical care has given greater prominence to nonmaleficence than to beneficence; however, the advent of modern science is changing that. The practice of medicine has long been dictated by the use of knowledge acquired in labs, tested in clinics, and verified by statistical methods. For many, the rapid advances in medical technology, including simulation, have produced great uncertainty as to *what is beneficial* and *what is least harmful*. In the article "Ethical Issues in the Application of Virtual Reality to Medicine," Whalley suggested that virtual reality training would lead to placing patients in virtual environments [15]. The concern at the time of his writing was that the research would be ahead of the interests of the

---

[†]The Turing test is an informal validation method well suited to validating models of human behavior, first proposed as a means to evaluate the intelligence of a computer system. As conventionally formulated, a computer system is said to be intelligent if an observer cannot reliably distinguish between system- and human-generated behavior at a rate better than chance. When applied to the validation of human behavior models, the model is said to have passed the Turing Test and to be valid if expert observers cannot reliably distinguish between model- and human-generated behavior. Because the characteristic of the system-generated behavior being assessed is its degree of indistinguishability from human-generated behavior, this test is clearly directly relevant to the assessment of the realism of algorithmically generated behavior, perhaps even more so than to intelligence, as Turing originally proposed. For more information, see "Verification and Validation" by Mikel D. Petty, in *Principles of Modeling and Simulation: A Multidisciplinary Approach* edited by John A. Sokolowski and Catherine M. Banks, Wiley, Hoboken, NJ, 2009.

patient, and that the technology could be prone to error that would introduce specific distortions. But simulation has moved forward and has made great strides in creating human representations and environments that have addressed those distortions. As a result, M&S is becoming an acceptable part of medical training.

Despite having its critics, simulation is at the center of classroom instruction as a result of the confluence of technological advancement and the dire shortages of trained personnel in the health care field. Teaching programs have seen numerous unfilled faculty positions because health professionals with advanced training (the pool of eligible educators) can earn more in clinical practice. Additionally, the practice of hands-on training in the health professions may not be sustainable as the predominant model for preparing health professionals [8]. Thus, simulation-based training may be the sensible and most efficient method of avoiding dangerous shortages of caregivers. Advocates of simulation in health care education view simulation as *augmenting training*; they contend that simulation does not need to provide a perfect representation of a real medical problem as long as the approximation of reality produces the educational outcome desired. The supporters of simulation-based training conclude that the effectiveness of simulation training is judged by how well students learn the relevant lessons and skills, not how well the simulation resembles a human being with a medical problem [8].

At the Center of the Question . . .

At the center of the ethical concerns about simulation in the MHS are the three key actors: the at-large community of *modelers*, who select (or are given) data and make assumptions as to the content and characterization of the model; the *simulationists*, who design the experiment (or training scenario); and the *analysts*, who interpret the results of the simulation. It is important to note that the modeler is dependent on data to develop his or her model and that the best data in any model design come from subject matter experts. Thus, models representing human behavior, such as decision making or human interaction, rely on the qualitative analysis of subject matter experts who study human behavior, such as social scientists. Representing human systems requires expertise in the study of the anatomy from those among the MHSs.

Simulationists need to know the parameters and capability of a model for the creation of an experiment. Thus, a simulationist's skill set should include an understanding of modeling with subject matter expertise. Analysts (who are often the users) interpret the results of the simulation and make use of it in prescribing a response. Analysts should have subject matter expertise as well as an understanding of how the model was developed, including its design, intent, and limitations. Critical to model creation and simulation development is the ability to obtain dependable results. Therein lies the crux of the ethical issue in representing human behavior and human systems—the development of techniques to firm-up soft data and the representation of those data.

In the article "Models, Measurement, and Computer Simulation: The Changing Face of Experimentation," Margaret Morrison argues that computer simulations have the same epistemic status as experimental measurement when looking at the role that models play in experimental activity, particularly measurement [16]. Morrison attributes *models as measuring instruments* and *simulation as the experimental activities*. A model can be based on some theoretical belief about the world that is suggested by the data, or it can sometimes be understood as simply a statistical summary of a data set. With that, data assimilation is an example that extends beyond straightforward issues of description, and models fill the gaps in observational data. Again, there is an emphasis on the need to represent soft data accurately because the knowledge associated with the measurement comes via the model [16]. Moving to the next step, simulation development, Gilbert and Troitzsch suggest that the major difference between simulation and experiment is that in the latter, one is controlling the actual object of interest, whereas in a simulation, one is experimenting with a model rather than with the phenomenon itself. They contend that computer simulation is similar to experimentation in that it starts with a mathematical model of the target system and application of discretizing approximations, which replaces continuous variables and differential equations with values and algebraic equations [17].

There is a school of thought that adheres to the principle that *models mediate between us and the world* and *between us and theories* by acting as objects of inquiry and the source of mediated knowledge of physical phenomena. This mediated knowledge is characteristic of the type of knowledge acquired in measurement contexts. Morrison contends that the connection between models and measurement is what provides the basis for treating certain types of simulation outputs as epistemically (cognitively) on a par with experimental measurements. As such, computer simulation enables representation of the evolution of a system to measure the values of specific properties as the system approaches a critical point [16].

There is also the question of *complicated systems versus complex systems.* How do they differ? These systems diverge based on the level of understanding of the system; for examples, a human system may have few parts, but it is complex because it is difficult to ascertain absolutes in the data, as human systems data are organic and dynamic. Thus, one cannot predict the behavior of a human system with any certainty. On the other hand, a finite element model or physics-based model may be complicated, due to its numerous parts, but it is not complex, in that it is predictable and because the data to model such a system are not soft data.

Modeling human systems is also challenging. Still, it is becoming more and more necessary in light of the need to use simulation to train health care providers, due to the known and predicted shortages in the health care community. As discussed above, the MHSs are making greater use of simulation for training and patient care. Simulation training tools enable health care professionals to sharpen their assessment and decision-making skills without risk to patients in realistic, challenging, immersive environments that are instrumented to provide meaningful performance feedback. Some simulation tools facilitate training-to-proficiency in error-prone tasks. There are also manikins that simulate breathing, exhibit pulse and mimic vital signs, and respond to treatment. This type of training accelerates traditional training methods of watching procedures, then practicing what has been observed. Advances in technology have introduced training environments that go beyond simply simulating the vital signs; in fact, some are capable of representing every orifice as naturally as possible, with the ability to spurt blood and mimic the effects of biological attacks. Researchers have also developed better representations of skin, blood, and bone so that wounds have the correct smell, feel, and physiological accuracy. Amazingly, these advances have come without the benefit of a closer relationship between the developer and user communities.

CONCLUSION

In this chapter we have introduced the fundamentals of M&S in the medical and health sciences, such as the modes of M&S (live, virtual, and constructive), to conceptual model design and the evolution of model development as a mathematical or physical representation. We also highlighted the various roles, uses, and needs of the developer and user populations. With a model in hand, simulation is introduced; and it is used to facilitate the reproduction of experimental, clinical, or educational data.

We drew attention to the need for a more uniform dialogue of simulation within and between the M&S engineering and M&S medical and health sciences domains. As such, the information is presented with these two student bodies in mind, to enlighten them as to what they are required to master so that they can dialogue and collaborate with an informed understanding.

A strong case endorsing the use of M&S in education and training is being made across institutions and research centers—the sheer need for a larger body of health care professionals leads that discussion. Still, there are some who question the credibility of M&S in this domain. An intriguing seminal piece entitled "Posthuman Future: Consequences of the Biotechnology Revolution," by Francis Fukuyama (Paul H. Nitze School of Advanced International Studies, Johns Hopkins University) cautions against blindly wielding the new tools that science creates. Fukuyama counsels developers and users to consider carefully each step taken in the name of health progress, and to guide the medical community with intelligent debate on the consequences of human intervention [6]. We acknowledge his concerns.

To mitigate the *blind wielding* of these new tools, in this chapter and the ensuing discussion in Chapters 2 through 10 we endeavor to build that much needed two-directional bridge where medicine meets virtual reality (a.k.a. modeling and simulation) and where modeling and simulation meets medicine.

KEY TERMS

clinical engineer	constructive mode	CIMIT
biomedical engineer	computational model	CMS
engineered device	physical model	VMASC
biomechatronics	humans as models	surrogacy
robotics	human systems	game-based learning
training simulation	modeling	haptic device
analysis	disease modeling	augmented reality
experimentation	virtual operating room	beneficence
training	standardized patient	nonmaleficence
modes of M&S	modified stethoscope	modeler
live mode	science of simulation	simulationist
virtual mode	AIMS	analyst

REFERENCES

[1] Sokolowski JA, Banks CM. A proposed approach to modeling and simulation education for the medical and health sciences. In *Proceedings of the 2010 Summer Simulation Conference*, Ottawa, Ontario, Canada, July 11–15, 2010.

[2] Begg R, Lai DT, Palaniswami M. *Computational Intelligence in Biomedical Engineering*. Boca Raton, FL: CRC Press, 2008.

[3] Bronzino JD, Ed. *Medical Devices and Systems*. Philadelphia: Taylor & Francis, 2006.

[4] Westwood JD, et al., Eds. *Medicine Meets Virtual Reality: Parallel, Combinatorial, Convergent NextMed by Design*. Amsterdam: IOS Press, 2008.

[5] Westwood JD, et al., Eds. *Medicine Meets Virtual Reality: Building a Better You, the Next Tools for Medical Education, Diagnosis, and Care*. Amsterdam: IOS Press, 2004.

[6] Westwood JD, et al., Eds. *Medicine Meets Virtual Reality: Next Med Health Horizon*. Amsterdam: IOS Press, 2003.

[7] Kyle RR, Murray WB, Eds. *Clinical Simulation: Operations, Engineering, and Management*. Amsterdam: Elsevier, 2008.

[8] Bauer JC. *The Future of Medical Simulation: New Foundations for Education and Clinical Practice*. White Paper for Technology Early Warning System, Jan. 2006. http://www.jeffbauerphd.com/TEWSMedicalSimulation.pdf. Accessed Dec. 16, 2009.

[9] Crookall D, Zhou M. Medical and health care simulation: symposium overview. *Simulation Gaming*, 32:142, 2001. http://sag.sagepyb.com/cgi/content/abstract/32/2/142. Accessed Dec. 16, 2009.

[10] Combs D. Modeling and simulation: real world examples. In *Principles of Modeling and Simulation: A Multidisciplinary Approach*, Sokolowski JA, Banks CM, Eds. Hoboken, NJ: Wiley, 2009.

[11] Riley RH, Ed. *Manual of Simulation in Health-Care*. Oxford, UK: Oxford University Press, 2008.

[12] McGaghie WC, Issenberg SB, Petrusca ER. Editorial—Simulation: savior or Satan? A rebuttal. *Adv Health Sci Educ*, 8:97–103, 2003.

[13] Norman G. Simulation: savior or Satan. *Adv Health Sci Educ*, 8:1–3, 2003.

[14] Turing AM. Computing machinery and intelligence. *Mind*, 59(236):433–460, 1950.

[15] Whalley LJ. Ethical issues in the application of virtual reality to medicine. *Comput Biol Med*, 25(2):107–114, 1995.

[16] Morrison M. Models, measurement, and computer simulation: the changing face of experimentation. *Philos Stud*, 143:33–57, 2009. DOI 10:1007/sl 1098-008-9317-y.

[17] Gilbert N, Troitzsch KG. *Simulation for the Social Scientist*. Philadelphia: Open University Press, 1999.

2 The Practice of Modeling and Simulation: Tools of the Trade

JOHN A. SOKOLOWSKI

INTRODUCTION

This chapter could be subtitled: "What Medical and Health Care Professionals Should Know about M&S." The aim is to provide these professionals with an overview of the core elements of modeling and simulation (M&S) so that they can better communicate and interface with those who develop models and simulations for their use. The obvious dilemma is that most medical and health care professionals (MHPs) are not trained in this discipline. Conversely, M&S professionals are often not familiar with the medical and health care professions. However, for effective development of M&S tools, each group must have some appreciation for what information must be exchanged to support tool development successfully. It is from this precept that we proceed.

MODELING AND SIMULATION TERMS AND DEFINITIONS

Most professions have a set of terms with specific definitions commonly understood by those in the profession. This statement is certainly true for those in the M&S discipline. To ensure that this common understanding exists between MHPs and those in the M&S profession, in this section we lay out the key terms that will form the shared vocabulary. We will begin with the concept of a model. In the M&S profession a **model** in its most basic form is understood to be *an abstraction of reality*. This abstraction could be of a physical object or could be some intangible concept. Many people mistake models as representing physical objects only, but a model can describe a social theory, an element of human behavior, or a relationship

Modeling and Simulation in the Medical and Health Sciences, First Edition. Edited by John A. Sokolowski and Catherine M. Banks.
© 2011 John Wiley & Sons, Inc. Published 2011 by John Wiley & Sons, Inc.

among statistical variables. As an example, Gary Klein developed a model of how human beings make decisions, especially if they have significant experience in a particular field or domain. His model, known as **recognition-primed decision making** (RPD), did not represent a physical object but a sequence of steps that he purports take place inside the mind when a person is faced with having to make a decision [1].

Various models for MHPs have been developed. Examples of these include organs made of rubber or plastic used in schools or offices for teaching or educational purposes, mathematical representations of biological processes found in the body, and manikins used to teach certain surgical procedures. Because of improvements in computer hardware and software, computer-generated models have emerged. This technology offers an added dimension to models supporting MHPs. Computer-generated models provide an important means of visualizing the human body and provide training environments that immerse MHPs in virtual worlds representing situations that mimic real-world environments. Examples of these are provided in later chapters.

By their very nature, models are static representations of the entities or concepts they represent. To depict how models behave over time, one must implement them in a simulation. Thus, the term **simulation** is defined as *models that have been implemented in a temporal manner*. The implementation may take one of three forms: *live, virtual, and constructive simulation*. Live simulation may seem to be simply a facet of the real world, but there is a distinction. From the M&S perspective, **live simulation** consists of real people using real equipment but employing the equipment outside the context of a real-world event. An example of live simulation is an actor portraying a patient for training purposes. An example of a live simulation built around actors known as *standardized patients* is provided in Chapter 5.

One can think of **virtual simulation** as consisting of real people employing simulated equipment. The best known examples of virtual simulation are the flight simulators used by airlines and the military for pilot training. Here real pilots train on cockpits mocked up to represent a specific type of aircraft cabin. Computers provide inputs to the flight instruments to replicate aircraft response as the pilot manipulates the controls. In the medical area, virtual simulation usually takes the form of manikins configured to facilitate a specific medical training objective. Probably the best known and most frequently used virtual medical simulation is *Resusci-Anne*, a manikin used to train students in cardiopulmonary resuscitation [2].

Constructive simulation involves simulated people working with simulated systems. The concept of constructive simulation may be somewhat difficult to grasp because human beings do interface with this simulation, but primarily from an observer standpoint. They may at times provide input to the simulation to cause it to go in a specific direction, but the simulation then actually carries out the action. The first-person shooter games that are popular among many computer game enthusiasts are examples of constructive simulation. Here the human being interacts with the game in a role-playing mode, but the computer actually performs the actions in the form of a virtual person immersed in a virtual environment.

These simulation forms do not need to exist in isolation. They can be combined, and often are, to produce a simulation environment known as *live–virtual–constructive* (LVC) *simulation*. One such example is the **virtual operating room** (VOR) developed by researchers at Old Dominion University and Eastern Virginia Medical School [3]. The VOR combines a manikin (virtual simulation) configured for gallbladder surgery with an immersive three-walled **CAVE** environment (constructive simulation).[†] The CAVE has avatars that represent the anesthesiologist, the circulating nurse, and various operating room (OR) equipment, all computer generated. The avatars have a degree of artificial intelligence that permits them to respond to a live person practicing gallbladder surgery just as their real counterparts would respond. This type of environment allows one to immerse the medical trainee in a world very closely representing what he or she will see in an actual OR and to pose stressful situations that could not be practiced by any other means.

Three other terms often used by modeling and simulation professionals are *verification, validation, and accreditation* (VV&A). They are important aspects of any M&S development effort and are essential prerequisites for the credible and reliable use of the M&S product. To illustrate, consider the following scenario utilizing the VOR concept noted above. Suppose that a surgeon desires to be certified to perform a specific type of surgical procedure. He or she may train on that procedure using the VOR simulation environment. After completing the surgery successfully a given number of times on the simulator, the surgeon may then be certified to perform it on a human being with no further supervision. The assumption here is that the VOR simulation is sufficiently accurate with respect to its real-world counterpart that performance of the surgery on a person under the supervision of a qualified surgeon is not necessary. This accuracy is assessed and assured through the VV&A process.

Although verification and validation may seem to have identical definitions, they are significantly different from one another in the context of modeling and simulation. **Verification** refers to how well a model's implementation conforms to its design specifications. The modeling question typically answered during the verification process is: Does the software code of the executable

---

[†]A CAVE is a virtual environment that surrounds a person on three or four sides, and possibly the floor and ceiling, with computer screens that have projected images giving the person the impression that he or she is actually in the area being represented in the computer rendering.

model correctly implement the conceptual model? In other words: Did I build the model right?

Validation answers a different question. It characterizes the degree to which the model is an accurate description of the real-world object or process that it was meant to represent. In comparison to verification, the complementary question for validation is: Did I build the right model?

Arguably, this accuracy criterion can be highly subjective, especially when dealing with concepts that are not readily quantifiable. Accuracy must be judged in the context of a model's intended use. One would expect models intended to teach delicate surgical techniques to be highly accurate in their representation of the real patient, whereas a model used to educate nonmedical or health care professionals on some aspect of anatomy or surgical process might not require the same degree of accuracy.

Accreditation is the third term in this area of modeling and simulation, but it is an entirely different process from the other two terms. Verification and validation are clearly testing processes that are technical in nature. Accreditation is more of a decision process and may be nontechnically based but could be informed by technical data. Formally, accreditation is the certification by a responsible authority that a model or simulation is acceptable for a particular use. Implied in this definition is the fact that accreditation is always for a specific use, although the use may be broad. As an example, a surgical simulation replicating childbirth may have completed the verification and validation process set up for its procurement. However, its use at a specific medical school or hospital would probably be decided or accredited by the governing body for that institution. There are a variety of accepted verification and validation techniques, which are beyond the scope of this chapter. The reader is referred to an article by Petty for an introduction to these techniques [4].

MODELING AND SIMULATION PARADIGMS

MHPs may have encountered live and virtual simulation in their professional careers. The standardized patient concept and virtual patient representation via manikins are not new concepts, the former conceived in the 1950s and the latter developed in the 1960s. What may be unfamiliar to this body of professionals is constructive simulation. These simulations are the computer-generated environments similar to computer games that immerse the player in a virtual environment strictly controlled by the computer. Two general modeling and simulation paradigms are employed to create constructive simulations. These paradigms are presented here to provide the reader with a basic introduction to their concepts and characteristics. They are often characterized from both a modeling perspective and a simulation perspective because of the blurring of these two concepts by non-M&S professionals and even by some of those professionals directly engaged in M&S. These paradigms are most often associated with simulation because of the inclusion of a temporal component and are presented here from that perspective. The intent of describing these paradigms is to provide the reader with an understanding of the concepts behind developing constructive simulations.

Discrete Event Simulation

The first such paradigm to explore is that of **discrete event simulation**. This paradigm relies on the occurrence of specific events to advance a simulation from one state to another, over time. Formally, it is defined as the variation in a model caused by a chronological sequence of events acting on it. *Events* are instantaneous occurrences that may cause variations or changes in the state of a system. The *state of a system* is defined as one or more variables that describe a system completely at any given moment in time. These variables are known as *state variables*. Here *system* refers to the entire collection of objects being simulated and their relationships to one another.

Almost everyone is familiar with the operation of traffic lights which control the movement of vehicles and pedestrians at roadway intersections. Discrete event simulation would be an appropriate choice of an M&S paradigm to use to develop a constructive simulation of such a light's operation. The light can be in one of three states: red, yellow, or green, so a state variable called *light color* could be chosen to represent its state. The value of this variable at any point in time completely describes the condition of the light. Events that could cause the light to change state (i.e., transition from green to yellow, or yellow to red, or red to green) may include the passage of a specific amount of time, or the sensing of the presence of a vehicle by an external sensor such as a camera or magnetic coil. Hence, the simulated traffic light will change state when one of these triggering events occurs. The point here is that these are discrete events that cause a change in the state of the simulated light.

A more complex example of discrete event simulation is that of a simulation mimicking what takes place on a daily basis in a hospital's emergency department (ED). In an ED, patients arrive at random times with various maladies. They require different services based on their particular illness, each of which has a different length of time to complete. State variables describing this system might include the number of patients in triage, the number of patients in exam rooms, and the number of patients awaiting lab work or x-rays. Events that trigger changes in the system state include the arrival of patients at the ED, discharging a patient when treatment is complete, completion of lab work, or referral of a patient to another portion of the hospital. Inside the constructive simulation an event queue is generated that represents and schedules these events. The schedule is derived from data used to develop statistical distributions of the arrival and processing times expected for patients in each part of the ED. Thus, many days of ED care can be played out in a very short period of time to understand the efficiency and possible problem areas affecting patient care. The simulation could be used to explore different patient processing policies to see what effect those policies would have on the overall functioning of the system. Thus, this constructive simulation becomes a tool to investigate process improvement when it is not acceptable to test concepts in the real-world system. For examples of how one can implement a discrete event simulation using common spreadsheets, see an article by Sokolowski [5].

Continuous Simulation

Continuous simulation differs from discrete event simulation in that the system represented in the simulation changes constantly over time, as opposed to changing state by some triggering event. Continuous simulation is most often encountered when one models phenomena governed by the laws of physics. Examples include mechanical, electrical, and thermal systems. A simple illustration would be the simulation of a ball dropped from a high elevation. Here gravity pulls the ball back to Earth. The ball is in continuous motion with increasing velocity.

In the medical and health care area, disease spread is often simulated using a continuous simulation method. A well-known underlying model for this simulation is the SIR (susceptible, infected, and recovered) model. These three segments encompass the status of a population at any point in time, with some portion of the population susceptible to the disease, some infected, and some that have recovered from the illness. In most simulations of this phenomenon, the population is not modeled individually. Instead, the rate of movement from one segment to another is represented as a continuous function, with each segment being monitored to see how the disease is progressing through the population. Figure 2.1 shows a plot of the disease cycle implemented with a continuous simulation. Summing the number of people in each of these segments at any given time provides the total population represented in the simulation. See an article by Roberts [6] for a discussion of the mathematical model behind this simulation.

Other Simulation Paradigms

The reader may come across simulations based on two other simulation paradigms: system dynamics and agent-based simulations. **System dynamics** simulation is essentially a continuous simulation that has a specific

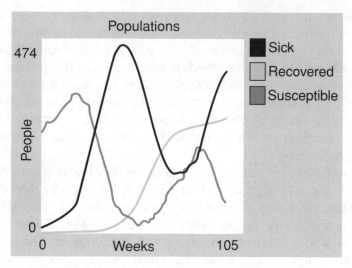

FIGURE 2.1   Continuous simulation of SIR disease cycle.

methodology associated with its development. It incorporates a formal method of constructing a model that represents the real-world system. This method identifies variables that govern the behavior of the system. It also identifies the relationships among those variables that produce the dynamic behavior of the system. Constructing a system dynamics simulation begins with a casual loop diagram. This diagram identifies the variables and their relationships responsible for system behavior. Figure 2.2 is an example of a causal loop diagram depicting the SIR model discussed above.

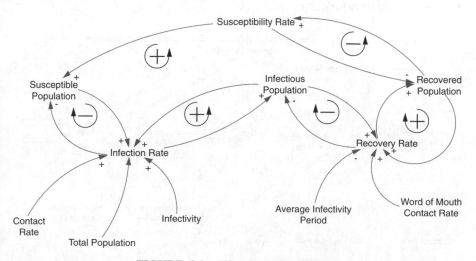

FIGURE 2.2   SIR causal loop diagram.

The arrows in Figure 2.2 depict the causal relationships between variables that define system behavior. The small plus signs (+) at the ends of the arrows indicate a positive causal effect (i.e., if the cause variable increases, the affected variable's value responds in the same way). It is just the opposite for negative relationships. The large plus and minus signs describe the loops that exist in the model. Loops are important because they indicate the various feedback paths that exist. Feedback is essential for the operation of system dynamics models since it helps define the overall response of the model to changing parameters.

One shortcoming of causal loop diagrams is their inability to directly represent the time rate of change of variables of interest. For this, one must translate the causal loop diagram into a stock and flow diagram, which captures the temporal aspect of the system. Figure 2.3 is a translation of the SIR model causal loop diagram into a stock and flow diagram. The rectangular blocks of this figure represent *stocks*, which are variables that one wishes to monitor. *Flows* are variables that feed into stocks directly and control the rate of change of the stock variable. Flows are represented by bowtie-looking symbols, which function as valves controlling the flow of data into and out of the stock. These flows can change over time, thus controlling how quickly the stocks change.

System dynamics models provide a method to represent conceptually and mathematically how a system behaves over time. The system can be anything from a physical entity such as a manufacturing plant to a process that people follow to accomplish a task. Figures 2.2 and 2.3 provide a visual representation of the system in a manner that allows one to see how variables are connected from a cause-and-effect standpoint, thus allowing a complex system to be viewed in a straightforward manner.

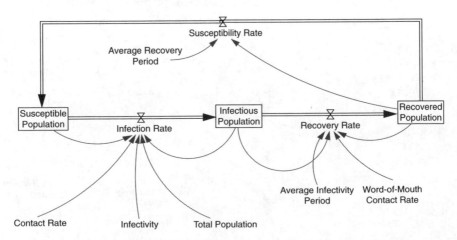

FIGURE 2.3 SIR stock and flow diagram.

Three population levels have been defined as the variables of interest for which level values are tracked. Those level values change as the result of the rate flows going into and out of the stocks. The rate values, in turn, are affected by other system variables that govern how those rates change. For a more detailed description of this simulation paradigm, see Sterman's treatment of the subject [7].

Agent-based simulations incorporate entities known as *software agents* to represent people, physical objects, processes, or concepts that contribute to the behavior of a system. This simulation paradigm differs from system dynamics in that system dynamics tends to take a top-down approach to simulating a system. It is concerned with how the system operates as a whole and not so much about the details of each part. Agent-based simulation takes more of a bottom-up approach to representing a system since it often captures detailed representations of individual parts of the system.

Agents have a unique definition within modeling and simulation. An *agent* is defined as an *autonomous software entity that senses its environment and acts to carry out a specific task or to achieve a specific goal.* The concept of autonomy is important in agent-based simulations. It ensures that all agents act based on their own set of beliefs or goals. Other agents or the environment can influence those actions but not direct them. When two or more agents are combined in simulation, autonomy leads to adaptive behavior of the system that was not predetermined by the modeler at the time of system design.

Autonomy can also introduce dilemmas for agents with conflicting goals. When two or more agents have conflicting goals, neither agent may be able to achieve its goals fully without some form of compromise. Agent-based simulations often have negotiation mechanisms built into them for this purpose. These mechanisms allow each agent to progress toward its goals in a manner acceptable to it. As one can tell from this discussion, an agent can behave very much like a person. Agents are often used to represent human behavior on an individual or group level.

As an example of an agent-based simulation, we revisit the SIR model. Recall that with the system dynamics representation of this disease model, we were modeling what was happening with the aggregate population in each of three categories: susceptible, infected, and recovered. The stocks of Figure 2.3 monitored the total number of people who fell into each of these groups as the disease spread through the population over time. A simulation of this disease spread could also be implemented using an agent-based simulation. Each agent could represent an individual member of a population or possibly a group of people in the population. The agent could be given specific immunity characteristics. It could interact with other agents on a specific schedule, much as people interact on a daily basis at work or school. Based on these interactions, one agent could pass on the disease to others, based on its incubation status and the other agent's immunity status. As one can see, this type of simulation can represent what is happening to the disease spread on an individual basis, thus allowing for a much more detailed view of disease transmission.

The simulation paradigm one chooses is really dependent on the purpose for which one is constructing the simulation. Is it needed for an aggregate assessment of an area, or is it to be used for a detailed investigation of a problem at a low level of representation?

CONCLUSION

In this chapter we introduce medical and health care professionals to the realm of modeling and simulation. We provide a common vocabulary for understanding common modeling and simulation terms and describe the major modeling and simulation paradigms that are used to both develop models and then to translate those modeling into working simulations. This chapter is by no means an exhaustive treatment of the scope of modeling and simulation but should provide the reader with enough familiarity to converse with professionals in this field when assisting in the development of models and simulations in support of health care.

KEY TERMS

model

recognized-primed
  decision making

simulation

live simulation

virtual simulation

constructive simulation

virtual operating room

CAVE

verification

validation

accreditation

discrete event simulation

continuous simulation

system dynamics

agent-based simulation

REFERENCES

[1] Klein G. *Sources of Power: How People Make Decisions*. Cambridge, MA: MIT Press, 1998.

[2] Cooper JB, Taqueti VR. A brief history of the development of mannequin simulators for clinical education and training. *Qual Saf Health Care*, 13(Suppl. 1): i11–i18, 2004.

[3] Emre B, Saurav M, Belfore LA. Simulation architecture for virtual operating room training. In *Proceedings of the 2009 Spring Simulation Multiconference*. San Diego, CA: Society for Computer Simulation International, 2009.

[4] Petty M. Verification and validation. In *Principles of Modeling and Simulation: A Multidisciplinary Approach*, Sokolowski JA, Banks CM, Eds. Hoboken, NJ: Wiley, 2009.

[5] Sokolowski JA. Simulation: models that vary over time. In *Principles of Modeling and Simulation: A Multidisciplinary Approach*, Sokolowski JA Banks CM, Eds. Hoboken, NJ: Wiley, 2009, pp. 47–69.

[6] Roberts MG. A Kermack–McKendrick model applied to an infectious disease in a natural population. *Math Med Biol J IMA*, 16:319–332, 1999.

[7] Sterman JD. *Business Dynamics: Systems Thinking and Modeling for a Complex World*. New York: McGraw-Hill/Irwin, 2000.

PART TWO
Modeling for the Medical
and Health Sciences

3 Mathematical Models of Tumor Growth and Wound Healing

JOHN A. ADAM

INTRODUCTION

Cancer is a complex phenomenon; to attempt to model any of its various facets (e.g., as avascular spheroid growth, angiogenesis and vascularization, invasion or metastasis) in a reasonable mathematical manner, many simplifications are necessary [2]. If the simplifications are reasonable, the model may be of considerable use, not only as a receptacle for what is already known, but also for its predictive capabilities. Frequently, however, the simplest mathematical models give considerable insight into complicated problems and suggest further improvements and directions for the next generation of model. The question may arise as to what constitutes a *model* as opposed to a metaphor (or even a simile) for cancer. This is well illustrated in the paper "Instability and Mitotic Patterns in Tissue Growth" by Glass, discussed below, which is very suggestive of the fundamental phenomenon—uncontrolled cellular proliferation—yet clearly is too simple to account for more complex aspects of tumor growth and development [3]. Several complementary levels of description are possible, and it is our purpose in this chapter to provide a brief survey of some of the basic mathematical models that have been developed over the past three decades. Further details may be found in the references, notably in reference [1].

An important question to be asked at the outset is: What *is* a **mathematical model?** One basic answer is that it is the formulation in mathematical terms of the assumptions and consequences believed to underlie a particular *real-world* problem. The aim of mathematical modeling is the practical application of mathematical models to help unravel the underlying mechanisms involved in biological (or other) processes. Common pitfalls include the

Modeling and Simulation in the Medical and Health Sciences, First Edition. Edited by John A. Sokolowski and Catherine M. Banks.

indiscriminate, naive, or uninformed use of models, but when developed and interpreted thoughtfully, mathematical models can (1) provide insight into the nature of the problem, (2) be useful in interpreting data, and (3) stimulate experiments. There is not necessarily a "right" model; obtaining results that are consistent with observations is only a first step and does not imply that the model is the only one that applies, or even that it is "correct." Furthermore, mathematical descriptions are not *explanations*, and never on their own can they provide a complete solution to a biological problem; often, complementary levels of description may be possible within the particular scientific paradigm. Collaboration with biologists (or whichever category of scientist is appropriate!) is needed for realism and help in modifying the model mechanisms to reflect the biology more accurately. On the other hand, workers in the biological sciences (for example) need to appreciate what mathematics (and its practitioners) can and cannot do! The mathematician needs to do the educating here; as always, good communication is necessary.

The art of good modeling relies on (1) a sound understanding and appreciation of the biological problem; (2) a realistic mathematical representation of the important biological phenomena; (3) finding useful solutions, preferably quantitative ones; and (4) biological interpretation of the mathematical results: insights, predictions, and so on. The mathematics is dictated by the biology and not, in general, vice versa, however tempting that may be. Sometimes the mathematics used can be very simple. The usefulness of a mathematical model should not be judged by the sophistication of the mathematics, but by different (and no less demanding) criteria [4]. It should also be pointed out that although the techniques of statistical analysis may frequently be used in portraying and interpreting data, the term *mathematical model* as used here refers to deterministic models as opposed to probabilistic or statistical ones, although many of the comments above also apply to these types of mathematical models. Indeed, as also pointed out by Gatenby and Maini, by contrast, a dominant theme of modern applied mathematics is that *simple underlying mechanisms may yield highly complex observable behaviors* [5]. This is the philosophical basis for the simple but suggestive models presented in this chapter. We concentrate here on deterministic models of tumor growth, with the exception of a computer simulation (based on a probabilistic model for tumor growth) presented by Williams and Bjerknes [6]. They defined tumor growth as occurring when a single abnormal cell divides faster than normal cells by a factor called the *carcinogenic advantage*. The simulations in two dimensions appear to have fractal-like boundaries, and this *metaphor* serves to define yet another complementary level of description in cancer biology [7].

It is to be hoped that mathematical models of tumor growth will eventually be able to incorporate as many of the following phenomena as possible (note that they are not mutually exclusive): mechanical and pressure effects, oxygen and nutrient distribution, growth inhibitor–activator distribution, destructive enzyme action, metabolic activity, blood vessel and capillary distribution, cell adhesiveness, the immune response, invasion, metastasis, and growth inhibition due to radiation, chemotherapy, or other treatment modalities.

At the simplest level of description, a population (of cells, bacteria, or people, depending on context) growing at a rate proportional to its present value increases in size exponentially. Historically, this is related to **Malthusian growth**, and is described by a simple differential equation if the continuum assumption is made. This assumption is that given a sufficiently large initial population, the rate of change of that population will be continuous in the standard mathematical sense. This assumption is not always appropriate, but when it is valid it has the pleasing consequence that differential and integral calculus may be used to describe many of the characteristics of that population's growth.

One can develop a sequence of models, increasing in mathematical complexity, which may describe the qualitative features of population growth in a given context. With enough free parameters in the model, it is possible to fit the growth curve to the empirical data and generate a reasonable fit. However, the questions to be asked of these models in the context of biological applications are: What are the biological mechanisms giving rise to the growth curve(s), and are they incorporated in the model? The simplest models are almost purely phenomenological in character and by themselves do not offer any biological insight into the nature of the problem, for example, in the study of tumor growth. As mentioned above, the simplest possible growth model (ignoring very trivial special cases) is that of exponential growth. Another simple model is that of limited growth: the population grows at a rate proportional to the *difference* between the present population and some limiting saturation value. This is a convenient model for describing the spread of information by mass media in a closed community, for example, or any other type of saturation phenomenon. A model that combines this saturation effect with the linear growth rate model mentioned above is the **logistic differential equations,** which incorporates both a linear growth rate term and a quadratic "self-interaction" term. This model has been used to describe a variety of growth situations, but it has the disadvantage that it is symmetric (i.e., the point of inflection occurs midway in the range of the function), which is atypical in practical terms. This problem can be overcome from an empirical point of view by generalizing the growth rate dN/dt in terms of a dimensionless parameter α [1]. A family of empirical curves of growth that saturates either more slowly or more rapidly than the logistic solution can be derived from a generalization of the Verhulst equation [1]. This formulation has the additional advantage that the logistic equation is recovered when $\alpha = 1$, and the Gompertz equation is recovered in the limit $\alpha \to 0$. The latter is useful in the modeling of some types of tumor in small animals [8].

OTHER BASIC MODELS OF TUMOR CELL POPULATION GROWTH: MULTICELLULAR SPHEROIDS

Other models have been developed by Marusic, Vaidya, and co-workers [9, 10]. Many **prevascular diffusion models** of tumor growth predict growth curves which are qualitatively similar to the saturated-growth curves derived from the models noted above, but a major advantage of these models is that the governing differential (or integrodifferential) equations of growth are based on plausible physical and biological assumptions. Thus, any comparison of model-generated curves of growth with experimental data can, at least in principle, provide some information on the appropriate parameter ranges necessary for consistency with the data.

As pointed out by Wheldon, the value of modeling to a science will depend on the extent to which that science incorporates defined assumptions that lend themselves to quantitative expression [11]. This is the basis of Greenspan's model, which gives the outer radius of the spheroid (and various inner radii, e.g., the radius of the necrotic core) as a function of time [12]. The first mathematical model we review, however, is that of Burton [13].

Burton was the first person to introduce diffusion into a model of spheroid growth, taking the model out of the realm of pure phenomenology into a more biologically driven environment. Interestingly, by considering a diffusion problem alone, Burton was able to glean important information about the relative thickness of the viable layer (the outer rim of proliferating cells) without utilizing an evolution equation for the outer tumor radius. The main features of models are easily summarized; the assumptions include spherical symmetry and diffusive equilibrium (see below for a discussion of this concept). The oxygen consumption rate per unit volume of nonnecrotic tissue is assumed to be constant, so the distribution of oxygen concentration satisfies a time-independent diffusion equation, subject to appropriate boundary conditions. The **necrotic core** is defined by a zero-flux boundary condition at the location where the oxygen concentration reaches a critical value below which cells are assumed to die. Further, at the outer surface of the spheroid the oxygen concentration is maintained at a constant value. Under these circumstances a simple mathematical relationship between these quantities is readily established. This simple model has obvious limitations: There is, in principle, no limit to the size of the spheroid (clearly, a major deficiency of the model!), but even here a useful result can be found for the relative size of the viable layer (the layer containing nonhypoxic cells, for example, wherein they still proliferate). It transpires that within the confines of this two-layer model, the viable layer approaches a constant thickness, equal to 58% of the critical tumor radius (i.e., the radius at which necrosis first occurs).

DIFFUSION OF GROWTH INHIBITOR

In 1973, Glass published a one-dimensional model of growth-inhibitor production in tissue; that model has proven to be seminal because of its simplicity and its usefulness. In a related fundamental paper, with more biologically realistic boundary conditions, Shymko and Glass dealt fully with the corresponding three-dimensional multicellular spheroid problem [14]. The former paper is summarized here, with the understanding that the basic biological features and consequences are not significantly different (apart from geometric factors) in the more realistic spherically symmetric case.

Consider a "slab" of slowly growing tissue of width L, centered on the origin (see Figure 3.1), producing growth inhibitor at a rate P (molecules per unit volume per second) that is depleted or decays at a rate λ, P and λ being constants. This central tissue is embedded in an infinite expanse of nonactive tissue. For simplicity, the diffusion coefficient D for the inhibitor is also assumed to be constant everywhere. A fundamental assumption made is that of diffusive equilibrium; the time scale for significant tissue growth is considered large compared with the time scale for readjustment of the inhibitor concentration profile due to such growth: The system is essentially in a steady state. This is a reasonable assumption for central tissue sizes not in excess

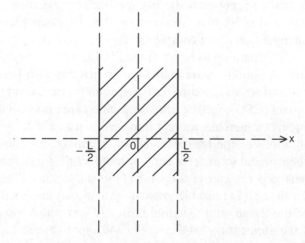

FIGURE 3.1   Basic configuration for Glass's model: an active tissue "swath" of width L embedded in an inert tissue plane.

of, say, 1 to 2 mm (typical of multicellular spheroids), for which the typical diffusion time (for oxygen) is on the order of 10 minutes [15].

The most straightforward boundary conditions of interest to us here are that the concentration of growth inhibitor C (in units of molecules per unit volume) should be smoothly varying across the boundary between the "active" tissue (producing the inhibitor) and the "normal" tissue, with the obvious symmetry condition that $C'(0) = 0$ (i.e., the derivative of the concentration of inhibitor is zero in the center of the slab; there is no flux there). The governing mathematical description can be represented in terms of an ordinary differential equation (rather than by a partial differential equation, in view of diffusive equilibrium).

Now we suppose, following Glass, that there exists a switchlike mechanism such that if the concentration C equals or exceeds some critical concentration θ, growth ceases in that region, and if $C < \theta$, growth continues. It follows that for growth to cease *throughout the tissue*, the concentration of growth inhibitor must satisfy $C \geq \theta$ at the edges (or boundaries) of the slab (or at the surface of the spheroid in the three-dimensional case). In terms of the dimensionless parameter $n = P/\lambda\theta$, it can readily be shown that the stable limiting size of tissue is given by the formula $L_s = \sqrt{D/\lambda} \ln [n/(n-1)]$. L_s is clearly defined for $n > 1$; under these circumstances growth eventually ceases as the system size approaches L_s. Glass speculated that the "singular region" defined in parameter space by $n \leq 1$ corresponds to the converse of controlled stable growth: namely, uncontrolled, unstable growth. This is a very basic definition of cancer: cellular proliferation defying the normal control mechanisms. Although claiming in essence to be no more than a metaphor for cancer at this level of description (given the lack of biological input insofar as cellular replication is concerned), this approach has been and continues to be an intriguing and suggestive avenue to a complex phenomenon that may well require many different but complementary levels of description to enable even a partial understanding to be obtained [7]. Details of the calculation are provided below. A similar mathematical approach has also been applied to the phenomenon of wound healing in bone, with particular reference to the critical size defect (CSD), further details of which are also provided below.

By introducing somewhat more biologically reasonable geometry and boundary conditions at the tissue boundary, Shymko and Glass were able to compare their model to experiments on multicellular spheroids [14]. The authors delineated parameter regions for (1) unstable (or unlimited) growth throughout the tissue, (2) unstable growth with mitosis confined to a peripheral region of the tissue, and (3) stable, limited growth. A very interesting discussion of the implications of the model and experimental data on cellular and geometric control of tissue growth is presented in their paper, and the reader is urged to consult it for further details. Subsequent developments of this model have focused on nonhomogeneous source terms and incorporation of necrosis as a possible source of inhibitor [16–19] (and references therein)].

TIME-EVOLUTIONARY DIFFUSION MODELS

The objective of the type of diffusion model developed by Greenspan is to infer the major internal process affecting tumor growth from the most easily obtained in vitro data, assumed to be (1) measurements of the outer nodule radius as a function of time, and (2) a cross section of the final dormant state, which provides the limiting radius and the limiting necrotic core radius of the final dormant in vitro state [12]. In this model there develops in general a three-layer structure consisting of a central necrotic core above which there is a layer of viable nonproliferating cells, and finally, an outer shell where all mitosis occurs. (No account is taken of the cell age or stage of cycle.) Greenspan lists many assumptions explicit or implicit in the model; the reader is referred to that paper for further details (see also [1], Chap. 2). It represents a fine example of mathematical modeling in biology. (The stability of such a spheroid, initially spherically symmetric, to nonsymmetric perturbations was considered elsewhere by Greenspan, and an extension is discussed below.)

The basis of a growth equation describing the evolution of the outer nodule radius can be described in the following manner: $A = B + C - D - E$, where A represents the total volume of living cells at any time t, B the initial volume of living cells at time $t = 0$, C the total volume of cells produced in $t > 0$, D the total volume of necrotic debris at time t, and E the total volume lost in the necrotic core for $t > 0$. One feature common to all these prevascular models is that the geometry is spherically symmetric; this is mathematically convenient, for in this ideal situation the net result of any intercellular forces within the colony is **radially directed**. Under these circumstances no more precise examination of local cell dynamics is needed. However, it is clear that in practice such spherical symmetry is likely to be very rare, and even if this is initially the case, subsequent growth of a small tumor cell colony is unlikely to maintain this symmetry. Chaplain and Sleeman [20] have contributed to the understanding of this situation, based in part on earlier work by Greenspan, and here we draw on the salient features of their work [20].

There is one caveat, and it applies to all models in which it is assumed that gross internal forces may be characterized by a pressure distribution, nonuniformities of which affect cell motion. Experiments on multicellular spheroids (R. K. Jain, 1995, private communication) failed to measure any pressure at all within the spheroids prior to vascularization. Once a spheroid or metastasis has been vascularized, the resulting pressure distribution must reflect the external systemic blood pressure and, by regarding the vasculariza-tion to correspond in some sense to a spatially smooth pressure distribution (as done by Adam and Noren [21]) the analysis in these models is still use-ful. Another prevalent assumption, also noted earlier, is that like molecular attraction, internal cell adhesion produces a "surface tension" at the outer boundary of the nodule that maintains compactness and counteracts internal expansive pressures. Although it is known that such multicellular spheroids are held together by a variety of junctions (see [20]), this postulated quasi-balance of forces (and associated surface area and volume considerations) is not sufficient to explain the stable limiting size of such a spheroid in vitro. Experimental work by Freyer [22] and others suggests that tumor growth inhibitors have at least some role to play in this regard. This is addressed to some extent by the work of Greenspan [12] (see also Greenspan, 1974) and its extensions (e.g., [19,23]).

With this in mind, we are in a better position to appreciate the strengths and weaknesses that characterize models of growing tumor colonies and their stability. An excellent introduction to models of this type is that of Jones and Sleeman [24], who identify the basic assumptions of Greenspan's model and rederive the governing equations. Thus formulated, the problem falls into the class known as *moving boundary value problems*, in which the outer surface is represented by the general expression $\Gamma(x,y,z,t) = 0$.

The interior pressure p and exterior relative nutrient concentration σ, re-spectively, satisfy the equations $\nabla^2 p = S$ inside $\Gamma = 0$ and $\nabla^2 \sigma = 0$ outside $\Gamma = 0$. S is a constant representing the rate of volume loss per unit volume within the cancer cell colony (due to necrosis, with the freely permeable necrotic debris being replaced by previously living cells via the compaction mechanism postulated above). Various other boundary and initial data are necessary to complete the statement of the mathematical problem. A prime concern of the model is **instability**: Do small perturbations to an initial equi-librium configuration continue to grow in time; or do they stabilize (as far as a linear analysis is concerned)? The simplest case mathematically is that of initial spherical symmetry, for which the equation $\Gamma(x,y,z,t) = 0$ becomes the simpler expression $r - R(t) = 0$, where $R(t)$ is the outer tumor radius and r is the distance of any interior point from the tumor center. Since the time variable enters the system only in determination of the tumor radius, the preceding system of equations can be solved explicitly. This enables a single ordinary differential equation for $R(t)$ to be derived. On solving this equation, it is found that the tumor (cell colony) typically grows in a logistic manner and asymptotes to the steady-state radius in a time scale regulated mainly by the rate of volume loss of necrotic debris.

Growth is judged unstable to infinitesimal perturbations if any such dis-turbance amplifies at an exponential rate [exceeding that of the radius $R(t)$].

In this circumstance, the instabilities radically alter the shape of the colony and can even lead to fracturing into two or more pieces. The tumor becomes unstable if and when it reaches a critical size beyond which surface tension is overcome by pressure forces. The precise criterion for this is obtained from the analyses of Greenspan (1976) and Chaplain and Sleeman [20] (see also reference [25]), who considered the growth of a spherical tumor (or cell colony) which is subject to small deviations that are always inevitably present in a real environment. Originally, for simplicity, Greenspan took the perturbations from complete sphericity to be axially symmetric [i.e., independent of the azimuthal angle φ in spherical polar coordinates (r, θ, φ)]. For completeness, Chaplain included the azimuthal dependence in his analysis. The equation of the moving surface is now written $r - R(t) - \varepsilon\eta(\theta, \varphi, t) = 0$. The solutions of this equation are given in terms of associated Legendre polynomials. It can be shown that an initial disturbance amplifies or decays according to whether a certain complicated expression is positive or negative. The tumor development is unstable if small perturbations can amplify; otherwise, it is stable. Under some circumstances the tumor is stable during its entire growth and attains its symmetric equilibrium configuration. It is possible, however, that the tumor becomes unstable at some definite time in its growth when small disturbances amplify and change both the structure and shape of the tumor. In this case, the changing pressure distribution overcomes the surface tension before spherical equilibrium is reached.

The onset of stability in the simplest nontrivial mode is manifested as a pinch in the outer surface of the tumor around the equatorial region. As the tumor grows and increases in size, other modes become unstable and more radical changes in configuration may occur, which can be used to model the tumor invading the surrounding tissues. Further work by Chaplain and Sleeman on tumor growth involves the use of results and techniques from nonlinear elasticity theory and differential geometry [26,27]. Their mathematical models describe the growth of a solid tumor using membrane and thick-shell theory. A central feature of their analysis is the characterization of the material composition of the model through the use of a strain-energy function, thus permitting a mathematical description of the degree of differentiation of the tumor explicitly in the model. Conditions are given in terms of the strain-energy function for the processes of invasion and metastasis, which frequently occur as the tumor evolves. These are interpreted as the modes of bifurcation of a spherical shell. The results are compared with actual experimental results and with the general behavior exhibited by both benign and malignant tumors. The authors also use their mathematical results in conjunction with aspects of surface morphogenesis in tumors (in particular, the Gaussian and mean curvatures of the surface of a solid tumor) in an attempt to produce a mathematical formulation and description of the important medical processes of staging and grading cancers.

TUMOR ANGIOGENESIS

Without the phenomenon of angiogenesis, solid tumors do not grow beyond a size of at most 2 mm in diameter (corresponding to approximately 1 million cells) in the absence of a blood supply. Indeed, this is the motivation behind the search for angiogenesis inhibitors: If this mechanism can be suppressed, a nodular carcinoma of this size is essentially harmless. Clearly, all the models discussed so far apply to the avascular or prevascular growth phase. The next level of biological and mathematical complexity requires an appropriate representation of vascularization, and to this end the work of Chaplain and Anderson (1996) is very important (see also a summary of this and other work in an article by Panetta et al. [28]. Their work focused attention on three features essential to the process: tumor angiogenic factors (e.g., vascular endothelial growth factor), endothelial cells, and matrix macromolecules (e.g., fibronectin).

The events leading to **angiogenesis** are as follows. At the end of the avascular phase the tumor cells secrete chemicals [collectively known as *tumor angiogenesis factors* (TAFs)] that induce endothelial cells in their vicinity to secrete matrixdegrading enzymes (e.g., proteases), which degrade the vascular basement membrane. These cells are then free to migrate through the membrane toward the tumor. Subsequently, the cells associated with the capillary sprout-tip region proliferate, leading to sprout growth, initially in parallel, but eventually they merge and form tip-to-tip and tip-to-sprout fusions and loop formations. Shortly after this, the first signs of circulation become evident, with new sprouts and buds repeatedly emerging, further extending the capillary network. Eventually, the tumor is penetrated, leading to vascularization and a vastly increased supply of oxygen for the tumor cells in the interior of the nodule.

The mathematical model developed by Chaplain and Anderson consists of three coupled partial differential equations describing the response of the endothelial cells to angiogenic factors via **chemotaxis** (whereby the cells are sensitive to and move up gradients of TAFs, from lower to higher concentrations) and to fibronectin via **haptotaxis** (fibronectin affects how endothelial cells adhere to collagen, and to the underlying substratum in general). Chaplain and Anderson solved the governing partial differential equations numerically (i.e., in discretized form) with parameter values based on experimental data. They were able to generate realistic capillary network structures, and by simulating the movement of endothelial cells toward a small circular tumor (in two dimensions, x and y), they found that several of the loops achieved anastomosis within 4.5 days and that complete vascularization was achieved at about 15 days.

Glass's Model

The governing differential equation is

$$\frac{\partial C}{\partial t} = D\frac{\partial^2 C}{\partial x^2} - \lambda C + PS(x) \tag{3.1}$$

where D, λ, and P are, respectively, the diffusion coefficient for the GF in the tissue, the decay or depletion rate of the GF, and the production rate of GF by the enhanced mitotically active cells in the vicinity of the wound edges, and the source term $S(x) = 1, |x| \leq L/2, S(x) = 0, |x| > L/2$ (also see Figure 3.1).

Under the diffusive approximation (the diffusion time scale is much smaller than a typical growth time scale), we may set the time derivative to zero. This assumption is valid for tumors on the order of a few centimeters or less in size, and is discussed in more detail in the section below on models of wound healing. The relevant boundary conditions are (1) $C'(0) = 0$, (2) $C(x)$ and $C'(x)$ are continuous at $|x| = L/2$, and (3) $\lim_{|x| \to \infty} C(x) = 0$. Now if we define the quantities $\alpha = (\lambda/D)^{1/2}$ and $\phi = \alpha L/2$, the differential equation (3.1) can be rearranged:

$$\frac{d^2C(x)}{dx^2} - \alpha^2 C(x) = -\frac{P}{D}S(x) \tag{3.2}$$

The solutions are as follows: (1)

$$C(x) = \frac{P}{\lambda}\left[1 - e^{-\phi}\cosh(\alpha x)\right] \qquad \text{for } |x| \leq \frac{L}{2} \tag{3.3}$$

and (2)

$$C(x) = \frac{P}{\lambda}e^{-\phi|x|}\sinh\phi \qquad \text{for } |x| > \frac{L}{2} \tag{3.4}$$

(See the generic graph of the solution in Figure 3.2.)

We note that $C(0) = (P/\lambda)\left(1 - e^{-\phi}\right)$ and $C(|L/2|) = P/2\lambda\left(1 - e^{-2\phi}\right)$ and that $C'(x) < 0$ for $x > 0$ (the graph being symmetric about the origin). If we recall that for growth to cease throughout $[-L/2, L/2]$, it is necessary that

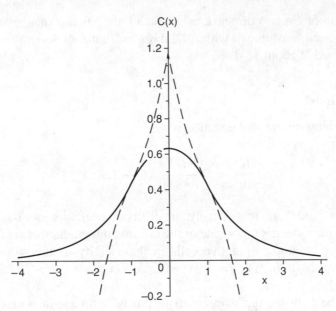

FIGURE 3.2 Complete solution (solid curve) for $C(x)$ based on equations (3.3) and (3.4). The dashed lines represent extensions of these solutions plotted outside their domains of validity to illustrate how the results (3.3) and (3.4) match across the tissue boundaries.

$C\,(|L/2|) \geq \theta$ [i.e., that $(P/2\lambda)\left(1 - e^{-2\phi}\right) \geq \theta$]. If we now form the dimensionless ratio $n = P/2\lambda\theta$, this condition is equivalent to $1 - e^{-2\phi} \geq n^{-1}$, or $L \leq L_s$, where $L_s = \sqrt{\lambda/D}\ln[n/(n-1)]$ is the *stable limiting size of the tissue*. As noted by Glass, it is tempting to identify the "occurrence of cancer" with the singular point $n = 1$ (i.e., $n \leq 1$ corresponds to limitless tissue growth) (see Figure 3.3). Thus, according to this model, a small change in the physiological parameters P, λ, *or* θ may be sufficient to switch a previously stable, well-regulated tissue system into one of (potentially) limitless growth (or, indeed, vice versa).

Time-Dependent Model

Consider the following "word equation" describing the rate of change of tumor volume:

$$\frac{d}{dt}\left(\begin{array}{c}\text{tumor}\\\text{volume}\end{array}\right) = \left(\begin{array}{c}\text{volume increase due}\\\text{to cell proliferation}\end{array}\right) - \left(\begin{array}{c}\text{volume decrease}\\\text{due to cell death}\end{array}\right) \quad (3.5)$$

With the following assumptions we can develop a simple model with some revealing characteristics. We assume that (1) the tumor is spherically symmetric at all times, (2) the rate of new cell production is proportional to the rate at which the tumor receives nutrient, (3) the rate at which the tumor receives

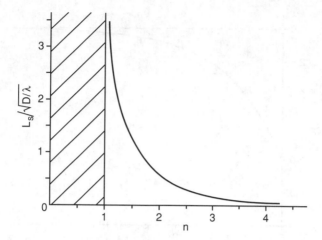

FIGURE 3.3 Stability boundary, demarking unstable tissue growth from stable tissue growth, according to the analysis of equations (3.3) and (3.4).

nutrient is proportional to the tumor surface area, and (4) the rate at which cells die (from lack of nutrient) is proportional to the tumor volume. Then, if $R = R(t)$ is the tumor radius, equation (3.5) can be written as

$$\frac{d}{dt}\left(\frac{4}{3}\pi R^3\right) = 4\pi k_1 R^2 - \frac{4}{3}\pi k_2 R^3$$

where k_1 and k_2 are constants of proportionality associated with nutrient release and absorption. Simplifying, we obtain the simple differential equation

$$\frac{dR}{dt} = k_1 - \frac{1}{3}k_2 R = k(K - R) \tag{3.6}$$

where $k = k_2/3$ and $K = 3k_1/k_2$. With the initial condition $R(0) = R_0$, a straightforward integration yields the solution

$$R(t) = K + (R_0 - K)e^{-kt} \tag{3.7}$$

Note that $\lim_{t\to\infty} R(t) = K$, the saturation level or carrying capacity. There-fore, if $R_0 < K$ (i.e., the initial tumor size is sufficiently small), the tumor grows to a limiting size K; this is determined, of course, by specific values of the constants k_1 and k_2, if known. If, on the other hand, $R_0 > K$ (the initial tumor size is sufficiently large), the tumor shrinks to the limiting size K. These solutions are illustrated in Figures 3.4 and 3.5, respectively. Note that increasing the value of k_1 (a measure of the rate at which tumor cells prolif-erate) and/or decreasing k_2 (a measure of the rate at which tumor cells die)

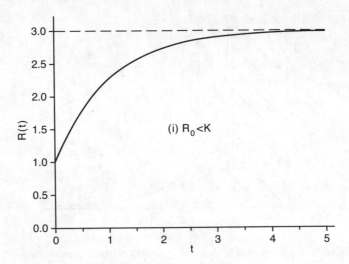

FIGURE 3.4   Growth curves for the simple time-dependent model based on equation (3.7): $R_0 < K$. (K here is chosen arbitrarily to be 3.)

will increase the limiting size K. Correspondingly, this size is decreased if k_1 is decreased and/or k_2 is increased. Generally, of course, these "constants" are both likely to be functions of the concentration of nutrient, which itself will depend on the location within the tumor. This spatial structure involves consideration of the diffusional aspects of the problem, discussed above, mathematical details of which may be found in reference [1].

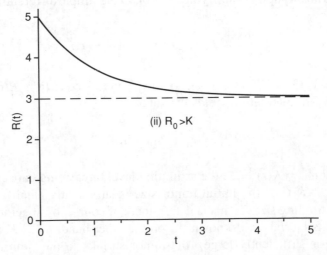

FIGURE 3.5   Growth curves for the simple time-dependent model based on equation (3.7): $R > K$. (K here is chosen arbitrarily to be 3.)

Wound Healing in Bone (With Particular Reference to the Critical Size Defect)

The fields of bone regeneration and wound healing in general often rely on suitable animal models to test experimental bone and tissue repair materials. One accepted model for the former is the **critical size defect** (CSD), which has been defined as the smallest intraosseous wound that does not heal by bone formation during the lifetime of the animal [29]. For practical purposes this time scale can usually be taken as one year. In the paper by Hollinger and Kleinschmidt [30], the definition was extended to a defect that has less than 10% bony regeneration during the lifetime of the animal. CSDs can "heal" by fibrous connective tissue formation, but since this is not bone, it does not have the properties (e.g., strength) that a completely healed defect would have. Some typical CSDs are, for rat, rabbit, dog, and monkey calvaria (skullcap), respectively: 8, 15, 20, and 15 mm (details may be found in an article by Schmitz and Hollinger [29]).

Wound healing, when it occurs, does so by means of a combination of various processes: chemotaxis (the movement of cells up a concentration gradient), neovascularization, synthesis of extracellular matrix proteins, and scar remodeling [31]. Growth factors are likely to play a very significant role in bone regeneration [32–35]. Such factors include transforming growth factor β, platelet-derived growth factor, insulinlike growth factor, and in the case of skin, epidermal growth factor [32,35]. Furthermore, the supply of oxygen to a wound has a strong influence on the quality of healing, and hence angiogenesis is of vital significance in bone and tissue regeneration [33,36].

In this section we summarize recent attempts to construct simple mathematical models of wound healing and bone regeneration, which reproduce some of the known qualitative features of those phenomena. Only one-dimensional Cartesian models are presented here, but this is easily generalized to the more realistic case of a planar circular wound (still technically one-dimensional if the only independent variable is the radius), and ultimately, to the case of a circular wound on a hemispherical surface. There is thus a sequence of models, each of which incorporates features not present in the models preceding. (Further details may be found in an article by Adam [37].)

It must be emphasized that the models discussed in this chapter do not address the time development of the wound in any way; they merely examine the conditions (e.g., wound size) under which such healing may occur. In these early models the primary focus is to account for the existence of a critical size defect by means of biochemical regulation of mitosis. Two related models can be examined although only the first is of direct interest for the CSD problem [38]. The first, and simpler, of the two corresponds to a "swath" of tissue (or more realistically, in this case, bone) removed from an infinite plane of tissue (see Figure 3.5) in which only a thin band of tissue at the wound edges takes part in tissue and bone regeneration. There is no tissue or bone in the interior. The second model has a geometric structure similar to that of the first, except that not all the tissue in the interior has been removed; it is a "gouge" or "graze" rather than a hole or puncture. This allows growth factor to diffuse across the wound, with consequent results for the existence of a CSD.

In these two models it is assumed that there is a thin layer of tissue (e.g., the epidermis) or bone (depending on the context) adjacent to the wound boundary that is responsible for increased mitotic activity at the edges of the wound by manufacturing a generic growth stimulator of concentration $C(x, t)$, where x is the direction of wound closure, and t is time, both in appropriate units discussed below. This layer or region is the spatial source of the growth factor production. $C(x,t)$, or its counterpart in other geometries, refers to the growth factor (GF) concentration which activates cell proliferation in the vicinity of the wound, and subsequent healing (partial or total).

Basic Configuration for Models IA and IB

We consider a one-dimensional wound of width L centered at the origin of coordinates (see Figure 3.6). At the wound edges, $x = \pm(L/2)$, as indicated above. We suppose that the generic growth factor is produced and that it is the distribution of this growth factor that determines whether or not wound healing occurs on the basis of this model. Before discussing the basic assumptions inherent in the model, we state the fundamental differential equation describing the space and time distribution of the growth factor concentration $C(x,t)$. It is given, as before, by

$$\frac{\partial C}{\partial t} - D\frac{\partial^2 C}{\partial x^2} + \lambda C = PS(x) \tag{3.8}$$

where, as before, D, λ, and P are, respectively, the diffusion coefficient for the GF in the tissue, the decay or depletion rate of the GF, and the production rate of GF by the enhanced mitotically active cells in the vicinity of the wound edges. These are all assumed to be constant in both models. Furthermore, $S(x)$ is the source term describing the distribution of GF production throughout the active tissue. In this model it is assumed to be uniform; thus, by symmetry, we can consider $x \geq 0$ without loss of generality, $S(x) = 1$, $L/2 \leq x \leq L/2 + \delta$, where δ is the thickness of the active layer; elsewhere, $S(x) = 0$. In equation (3.8) the first term represents the time rate of change of GF concentration, the second term describes the spatial change due to diffusion of GF, and the third term is the depletion or decay rate of GF as it interacts with the system as a whole, and is changed or removed. Thus, in the absence of diffusion and

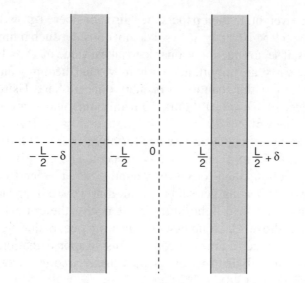

$$-\frac{L}{2}-\delta \qquad -\frac{L}{2} \qquad 0 \qquad \frac{L}{2} \qquad \frac{L}{2}+\delta$$

FIGURE 3.6 Basic configuration for wound healing models 1A and 1B: active wound edges of width δ embedded in inert tissue.

production, an initial distribution of GF will decay exponentially according to this equation.

Basic Assumptions

Several assumptions have already been noted, but in this section we identify the more important ones and their implications. The first to be noted is that of diffusive equilibrium, referred to earlier. Basically, this means that the process of readjustment of the GF concentration as the wound heals is so fast (compared with the typical wound-healing time) that, to a first approximation, the distribution of GF may be considered independent of time. This also simplifies the mathematics considerably! In to justify this assumption, consider the diffusion time scale t as defined from equation (3.1) or (3.8): $t \approx l^2/D$, where l refers to a typical length scale (size) of the system (i.e., the wound). The value of D depends, of course on the particular GF or enzyme in general (the higher the molecular weight, the smaller is D) and on the medium in which it is diffusing. However, some indication of this can be found by considering the diffusion of oxygen and sucrose in water. At a temperature of 25°C, $D \approx 2.4 \times 10^{-5} \mathrm{cm}^2 \mathrm{\ s}_{-1}$, while for sucrose at 20°C, $D \approx 4.6 \times 10^{-6} \mathrm{cm} \mathrm{\ s}^{-1}$ [15]. Sherratt and Murray [39] carried out a best-fit analysis from data on epidermal wound healing (there being no direct experimental data from which D could be determined) and estimated that for epidermal GF, $D \approx 3.1 \times 10^{-7} \mathrm{cm}^2 \mathrm{\ s}^{-1}$, considerably smaller because of the high molecular weight. In their papers they also considered growth inhibitors, for which $D \approx 5.9 \times 10^{-6} \text{cm}^2 \, \text{s}^{-1}$ (we do not consider such inhibitors in this chapter). Thus, it seems not unreasonable to take a value of $D \approx 10^{-5} \text{cm}^2 \, \text{s}^{-1}$ for oxygen (clearly an important factor in wound healing) and $D \approx 5 \times 10^{-7} \text{cm}^2 \, \text{s}^{-1}$ for GF, the quantity of primary concern here. Using this value of D for l-values of 1 μm (10^{-4} cm), 10 μm, 1 mm, and 1 cm, we find typical diffusion times of 2×10^{-2} s, 2 s, ≈ 5.5h, and ≈ 23 days, respectively. The corresponding diffusion time scales for oxygen, it should be noted, are 10^{-3} s, 10^{-1} s, ≈ 15 min, and ≈ 1 day, respectively. Clearly, the approximation is less well justified for GF in wound sizes of order 1 cm if we are considering wound healing per se, but recall that we are interested here in a mechanism that may shed light on the existence of the critical size defect: that wound size above which no essential healing occurs during the lifetime of the animal. Over such a time scale, the diffusive approximation is certainly a very good one for GF distributions. Under these circumstances, $\partial C/\partial t = 0$ in equations (3.1) and (3.8).

The second assumption is that the tissue growth or bone regeneration is regulated by the GF concentration $C(x)$ (recall: no time dependence for C, in light of the first assumption) via a discontinuous switch mechanism, such that increased mitotic activity and hence regeneration occurs at the wound edges when the GF concentration reaches or exceeds a critical or threshold value θ [i.e., when $C(\pm L/2) \geq 0$]. The third basic assumption is that there are no mechanical constraints; by this we mean that the tissue or bone is free to grow (when the criterion above is satisfied) into the wound space without any resistive pressure constraints (e.g., as would be present for an expanding benign tumor). Fourth, we make explicit an already implicit assumption: namely, that of the continuum approximation. This means that the dependent variable $C(x)$ is a continuous and suitably differentiable function; on the present scale of description we do not encounter the discontinuities that must inevitably be present on the molecular scale. We are now in a position to discuss model 1. Because in both models the system and solutions are symmetric about $x = 0$, we shall only address the domain $x > 0$. Corresponding results for $x < 0$ are then readily established.

Equations and Solutions: Model 1A

The governing differential equation may now be written in the simple form

$$\frac{d^2C}{dx^2} - \alpha^2 C = -\frac{P}{D}, \qquad \frac{L}{2} \leq x \leq \frac{L}{2} + \delta \qquad (3.9)$$

and

$$\frac{d^2C}{dx^2} - \alpha^2 C = 0, \qquad x > \frac{L}{2} + \delta \qquad (3.10)$$

[Recall that the domain of $C(x)$ is $(-\infty, -L/2] \cup [L/2, \infty)$.] Here the constant $\alpha = \sqrt{\lambda/D}$. The boundary conditions to be satisfied are:

1. $C(x)$, $C'(x)$ are both continuous at $x = L/2 + \delta$.
2. $\lim_{x \to \infty} C(x) = 0$.
3. $C'(L/2) = 0$.

The second of these conditions is necessary because there are no distant sources of GF production, so the concentration must decrease as the distance from the wound increases. The final condition means that there is no flux of GF into the (empty) interior. This will be modified in model 1B, for which interior tissue will be present. After some algebraic manipulation it follows that in the active or "epidermal layer" defined by $L/2 \le x \le L/2 + \delta$, the concentration of GF is given by

$$C(x) = \frac{P}{\lambda} \left\{ 1 - [\exp(-\alpha\delta)] \left[\cosh\alpha\left(x - \frac{L}{2}\right) \right] \right\} \qquad (3.11)$$

In the region exterior to the wound, $x \ge (L/2) + \delta$, the corresponding solution is

$$C(x) = \frac{P}{\lambda}(\sinh\alpha\delta)\left[\exp\alpha\left(\frac{L}{2} - x\right)\right] \qquad (3.12)$$

Using equation (3.11), we apply the criterion $L\,(|L/2|) \ge 0$, to obtain

$$\delta \ge \delta_c = \alpha^{-1}\ln\frac{n}{n-1}, \qquad n > 1 \qquad (3.13)$$

where n is a parameter defined in terms of the tissue constants P, λ, and θ by $n = P/\lambda\theta$. This clearly places, for given n, a lower bound (δ_c) on the thickness of the active layer necessary for the wound to heal. The region above the curve in Figure 3.7 corresponds to thicknesses δ for which healing and regeneration occur; below the curve no such event takes place—the active region is too thin to sustain the required level of GF production and retention.

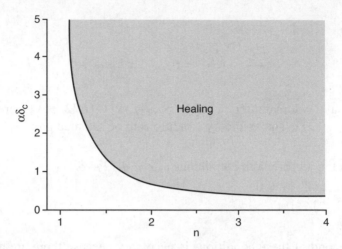

FIGURE 3.7 Healing boundary for model 1A: a measure of the wound edge thickness required for healing to occur as a function of the parameter n.

Estimates of Parameter Values

We have already noted some possible values for the diffusion coefficient D, but other quantities are still harder to pin down for a conceptual model of this type. Based on studies of DNA synthesis suppression by repeated injection of epidermal extract, Sherratt and Murray [39] estimated the half-life of chemical decay as 12 h, so for pure exponential decay, this corresponds to $\lambda = (\ln 2)/12 \text{ h}^{-1}$ or approximately $1.6 \times 10^{-5} \text{ s}^{-1}$. For $D \approx 5 \times 10^{-7} \text{cm}^2 \text{ s}^{-1}$ this gives $\alpha \approx 6 \text{ cm}^{-1}$. The most difficult of the parameters to assess is the ratio P/θ although this does not require that we know each quantity independently. The reciprocal of this ratio is a measure of how long it would take the active region to pump out enough GF to initiate the healing process (by reaching the threshold concentration θ) in the absence of GF decay and diffusion. With these processes included, of course, this takes considerably longer. It seems entirely reasonable to expect that $P << \theta$, so noting from model 1 the requirement that $n > 1$, we must have $P/\theta > 1.6 \times 10^{-5} \text{ s}^{-1}$. This means that even in the absence of the depletion effects mentioned above, the time required from wounding to the start of the healing process (not the time to heal, note) is at most about 17 h. This will be increased, of course, by the presence of depletion. We are now in a position to estimate the critical thickness of the active region δ by using equation (3.13). We write this now as

$$\delta_c \approx \frac{1}{6} \ln \frac{n}{n-1} \qquad (3.14)$$

For n in the range [1.2, 3.0] this expression places δ_c approximately in the range 0.30 to 0.07 cm. As noted above, the graph of the dimensionless quantity $\alpha\delta_c$ is shown in Figure 3.7, as a function of the parameter $n = P/\lambda\theta$. Note that $\alpha\delta_c$ is undefined for $0 \le n \le 1$. Healing occurs in the region lying above the curve.

Equations and Solutions: Model 1B

In the swath model, as indicated above, there is still some tissue in the wound interior (i.e., for $-L/2 \le x \le L/2$). However, it is considered to be dormant in that it does not contribute to the healing process. As before, the wound edges of thickness δ are the domains of GF production. As in model 1A, we will invoke spatial symmetry [i.e., $C(x) = C(-x)$] to allow the mathematical convenience of working with $x > 0$ only. The boundary conditions now are slightly different: We demand that $C(x)$ and $C'(x)$ are both continuous at $x = L/2$ and at $x = L/2 + \delta$, that $C'(0) = 0$ and as before, $\lim_{x \to \infty} C(x) = 0$. There are now three regions to consider for x. The governing differential equation is unchanged except that the homogeneous form now applies both in the wound interior $0 \le x \le L/2$ and exterior $x \ge L/2 + \delta$. In $0 \le x \le L/2$, the solution is

$$C(x) = \frac{P}{\lambda} e^{-\alpha L/2} \left(1 - e^{-\alpha\delta}\right) \cosh \alpha x \qquad (3.15)$$

In the active region of GF production, $L/2 \le x \le L/2 + \delta \equiv m$,

$$C(x) = \frac{P}{\lambda} (1 + A \cosh \alpha x + B \sinh \alpha x) \qquad (3.16)$$

where $A = -\left[\sinh(\alpha L/2) + e^{-\alpha m}\right]$ and $B = \sinh(\alpha L/2)$.

Note that the maximum value of GF concentration occurs in the active region, as would be expected. In the exterior $x \ge m$, the solution is

$$C(x) = \frac{PF}{\lambda} e^{-\alpha x} \qquad (3.17)$$

where $F = -e^{\alpha m} (A \sinh \alpha m + B \cosh \alpha m)$, and $m = (L/2) + \delta$. Of particular interest once again is the condition for healing at the wound edge: namely, $C(\pm L/2) \ge \theta$. From equation (3.14) this can be rearranged to yield an expression for the width of the wound such that it will heal. This inequality is

$$L \le L_c = \alpha^{-1} \ln \frac{N(\delta)}{2 - N(\delta)} \tag{3.18}$$

where $N(\delta) = n\left(1 - e^{-\alpha\delta}\right)$ and $n = P/\lambda\theta$, as before. Clearly, n is dependent on the active region thickness δ. Thus if L is below the critical width L_c defined by this expression, healing or regeneration occurs, and above this critical width it does not. The dimensionless quantity αL_c is illustrated in Figure 3.8. The mathematical restrictions on N are $1 < N(\delta) < 2$. Choosing a representative value of $\delta \approx 0.18$ cm and $n = 3$, we find from (3.16), using $\alpha = 6$, that $N \approx 1.98$, whence from (3.18), $L_c \approx 0.75$ cm; that is, *the critical size defect* is about 0.75 cm for this choice of parameters. A reasonable question may be asked at this stage: Why do we choose a value for δ of 0.18 cm rather than $\delta_c = 0.07$ as indicated for this choice of n? The answer is that the model indicates that δ is at least δ_c, so we are free to choose a larger value consistent with biological considerations. A further point to be noted is that models 1 and 2 are related in formulation, but are independent, so we use only general information on δ as provided by model 1 to ascertain general features from model 2, such as the size of L_c. It is also clear that some, indeed many, choices of parameter values will give very small (and hence unrealistic) values of the critical size defect. In our present state of knowledge about, for example, values of D, λ, and P/θ, and hence n, δ, and N, what we have been able to accomplish is to isolate parameter ranges that will give reasonable values for both the thickness of the

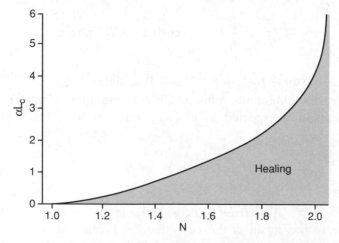

FIGURE 3.8 Healing boundary for model 1B: a measure of the wound size required for healing to occur as a function of the parameter *N*.

active region and the critical size defect, and also to establish that the models discussed here have the sensitivity to place reasonable bounds on such parameter values. Note that $0 < N < n$, these extremes being determined by the (unphysical) values of active region thickness δ, being zero and infinity, respectively.

CONCLUSION

A survey is provided of deterministic models of tumor growth and development over the last several decades and some recent models of wound healing. The evolution of these models has proceeded from basic phenomenological and empirical descriptions, through both time-dependent and time-independent diffusion models (largely within the diffusive equilibrium approximation). This includes a study of the diffusion of growth inhibitory and other factors. The linear stability of models of spheroids to asymmetric perturbations is discussed briefly, as are recent applications of nonlinear elasticity theory and differential geometry to possible staging and grading of cancers. A model of angiogenesis is summarized, illustrating the roles of tumor angiogenesis factors, endothelial cell migration, and matrix-degrading enzymes in the development of the tumor cell colony from an avascular nodule to a fully vascularized tumor. The underlying mathematical details for most of these models are not discussed here, but may be found readily in the original papers listed in the references. Several of the topics discussed here are discussed in Adam's article "General Aspects of Modeling Tumor Growth and Immune Response" [1]. Several simple mathematical models of tumor growth and wound healing are discussed, however, to illustrate the process of modeling in unusual contexts.

We have seen implicitly from the examples above that certain fundamental steps are necessary in developing a mathematical model (see Figure 3.9): formulating a real-world problem in mathematical terms using whatever appropriate simplifying assumptions may be necessary; solving the problem thus posed, or at least extracting sufficient information from it; and finally, interpreting the solution in the context of the original problem. Thus, the art of good modeling relies on (1) a sound understanding and appreciation of the problem; (2) a realistic, but not unnecessarily mathematical representation of the important phenomena; (3) finding useful solutions, preferably quantitative ones; and (4) interpretation of the mathematical results, yielding insights, predictions (such as when a tumor will shrink under the action of a radiation or chemotherapeutic regimen), and so on. Sometimes the mathematics used can be very simple, as above; indeed, the usefulness of a mathematical model should not be judged by the sophistication of the mathematics, but

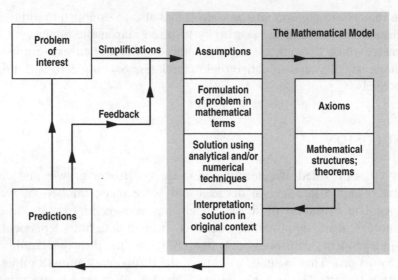

FIGURE 3.9   Flowchart illustrating the various components of a mathematical model. (From [40], by permission of Princeton University Press.)

by its predictive capability, among other factors. Mathematical models are not necessarily right (although they may be wrong as a result of ignoring fundamental processes). One model may be better than another in that it has better explanatory features, or more specific predictions can be made which are subsequently confirmed, at least to some degree. Sometimes models can be very controversial: This is a good thing, for it generates scientific and mathematical discussion.

Indeed, we can go further and suggest that all mathematical models are flawed to some extent: many by virtue of inappropriate assumptions made in formulating the model, or (which may amount to the same thing) by the omission of certain terms in the governing equations, or even by misinterpretation of the mathematical conclusions in the original context of the problem. Occasionally, models may be incorrect because of errors in the mathematical analysis, even if the underlying assumptions are valid. And paradoxically, it can happen that even a less accurate model is preferable to a more mathematically sophisticated one; it was the mathematical statistician John Tukey who stated that "it is better to have an approximate answer to the right question than an exact answer to the wrong one" [40].

It is fitting, therefore, to end with quotes from the article by Gatenby and Maini [5]: "Experimental oncology is awash with data. In 2001 alone, over 21,000 articles on characterizing, diagnosing and treating malignancies were published ... Remarkably, despite this wealth of information, clinical oncologists and tumour biologists possess virtually no comprehensive theoretical model to serve as a framework for understanding, organizing and applying these data . . . Mathematical models are typically denounced as 'too simplistic' for complex tumour-related phenomena (ignoring, of course, the fact that similar simplifying assumptions are required in most experimental designs) . . . Existing mathematical models may not be entirely correct. But they do represent the necessary next step beyond simple verbal reasoning and linear intuition. . . . [U]nderstanding the complex, non-linear systems in cancer biology will require ongoing interdisciplinary, interactive research in which mathematical models, informed by extant data . . . guide experimental design and interpretation."

KEY TERMS

mathematical model

Malthusian growth

logistical differential
   equation

prevascular diffusion
   model

necrotic core

radially directed

instability

angiogenesis

chemotaxis

haptotaxis

critical size defect

REFERENCES

[1] Adam JA. General aspects of modeling tumor growth and immune response. In *A Survey of Models for Tumor-Immune System Dynamics*, Adam JA, Bellomo N, Eds. Boston: Birkauser; 1997, pp. 15–87.

[2] Weinberg RA. *One Renegade Cell*. New York: Basic Books, 1998.

[3] Glass L. Instability and mitotic patterns in tissue growth. *J Dyn Syst Meas Control*, 95:324–327, 1973.

[4] Murray JD. On pattern formation mechanisms for lepidopteran wing patterns and mammalian coat markings. *Philos Trans R Soc (London)*, B295:473–496, 1981.

[5] Gatenby RA, Maini PK. Cancer summed up. *Nature*, 421:321, 2003.

[6] Williams T, Bjerknes R. Stochastic model for an abnormal clone spread through epithelial basal layer. *Nature*, 236:19–21, 1972.

[7] Adam JA. On complementary levels of description in applied mathematics: II. Mathematical models in cancer biology. *Int J Math Educ Sci Technol*, 19:519–535, 1988.

[8] Laird AK. Dynamics of tumor growth: comparison of growth rates and extrapolation of growth curve to one cell. *Br J Cancer*, 19:278–291, 1965.

[9] Marusic M, Bajzer Z, Vuk-Pavlovic S, Freyer JP. Tumor growth in-vivo and as multicellular spheroids compared by mathematical models. *Bull Math Biol*, 56:617–631, 1994.

[10] Vaidya VG, Alexandro FJ. Evaluation of some mathematical models for tumor growth. *Int J Bio-Med Comput*, 13:19–35, 1982.

[11] Wheldon TE. *Mathematical Models in Cancer Research*. Bristol, UK: Adam Hilger, 1988.

[12] Greenspan HP. Models for the growth of a solid tumor by diffusion. *Stud Appl Math*, 51:317–340, 1972.

[13] Burton AC. Rate of growth of solid tumors as a problem of diffusion. *Growth*, 30:159–176, 1966.

[14] Shymko RM, Glass L. Cellular and geometric control of tissue growth and mitotic instability. *J Theor Biol*, 63:355–374, 1976.

[15] Edelstein-Keshet L. *Mathematical Models in Biology*. New York: Random House, 1988.

[16] Adam JA. A simplified mathematical model of tumor growth. *Math Biosci*, 81:229–244, 1986.

[17] Adam, JA. A mathematical model of tumor growth: II. Effects of geometry and spatial non-uniformity on stability. *Math Biosci*, 86:183–211, 1987.

[18] Swan GW. The diffusion of inhibitor in a spherical tumor. *Math Biosci*, 108:75–79, 1992.

[19] Adam JA, Maggelakis, S. A mathematical model of tumor growth: IV. Effects of a necrotic core. *Math Biosci*, 97:121–136, 1989.

[20] Chaplain MAJ, Sleeman BD. Modelling the growth of solid tumors and incorporating a method for their classification using nonlinear elasticity theory. *J Math Biol*, 31:431–473, 1993.

[21] Adam JA, Noren, R. Equilibrium model of a vascularized spherical carcinoma with central necrosis: some properties of the solution. *J Math Biol*, 31:735–745, 1993.

[22] Freyer JP. The role of necrosis in regulating the growth saturation of multicellular spheroids. *Cancer Res*, 48:2432–2439, 1988.

[23] Adam JA. Maggelakis, SA. Diffusion regulated growth characteristics of a prevascular carcinoma. *Bull Math Biol*, 52:549–582, 1990.

[24] Jones DS, Sleeman BD. *Differential Equations and Mathematical Biology*. London: George Allen & Unwin, 1983.

[25] Byrne HM, Chaplain MAJ. Growth of non-necrotic tumors in the presence and absence of inhibitors. *Math Biosci*, 130:151–181, 1995.

[26] Chaplain MAJ, Sleeman BD. A mathematical model for the growth and classification of a solid tumor: a new approach via nonlinear elasticity theory using strain-energy functions, *Math Biosci*, 111:169–215, 1992.

[27] Chaplain MAJ. The development of a spatial pattern in a model for cancer growth. In *Experimental and Theoretical Advances in Biological Pattern Formation*, Othmer HG, Maini PK, Murray JD, Eds. New York: Plenum Press, 1993, pp. 45–60.

[28] Panetta JC, Chaplain MAJ, Adam JA. The mathematical modelling of cancer: a review. *In Mathematical Models in Medical and Health Science,* Horn MA, Simonett G, Webb, GF, Eds. Nashville, TN: Vanderbilt University Press, 1998, pp. 281–309.

[29] Schmitz JP, Hollinger JO. The critical size defect as an experimental model for craniomandibulofacial nonunions. *Clin Ortho Relat Res,* 205:299–308, 1986.

[30] Hollinger JO, Kleinschmidt JC. The critical size defect as an experimental model to test bone repair materials. *J Craniofac Surg,* 1:60–68, 1990.

[31] Bennett NT, Schultz GS. Growth factors and wound healing: biochemical properties of growth factors and their receptors. *Am J Surg,* 165:728–737, 1993.

[32] Mundy GR. Regulation of bone formation by bone morphogenetic proteins and other growth factors, *Clin Orthop,* 324:24–28, 1996.

[33] Marx RE,et al. Platelet-rich plasma growth factor enhancement for bonegrafts. *Oral Surg Oral Med Oral Pathol Oral Radiol Endodont,* 85:638–646, 1998.

[34] Nissen NN,et al. Vascular endothelial growth factor mediates angiogenic activity during the proliferative phase of wound healing. *Am J Pathol,* 152:1445–1452, 1998.

[35] Hsieh SC, Graves DT. Pulse application of platelet-derived growth factor enhances formation of a mineralizing matrix while continuous application is inhibitory, *J Cell Biochem,* 69:169–180, 1998.

[36] Eisinger M, Sadan S, Silver IA, Flick RB. Growth regulation of skin cells by epidermal cell–derived factors: implications for wound healing. *Proc Natl Acad Sci USA,* 85:1937–1941, 1988.

[37] Adam JA. Inside mathematical modeling: building models in the context of wound healing in bone. *Discrete Continuous Dyn Syst,* 4:1–24, 2004.

[38] Winet H. The role of microvasculature in normal and perturbed bone healing as revealed by intravital microscopy. *Bone,* 19:39S–57S, 1996.

[39] Sherratt JA, Murray JD. Mathematical analysis of a basic model for epidermal wound healing. *J Math Biol,* 29:389–404, 1991.

[40] Adam JA. *A Mathematical Nature Walk.* Princeton, NJ: Princeton University Press, 2009.

FURTHER READING

Adam JA. Mathematical models of prevascular spheroid development and catastrophe-theoretic description of rapid metastatic growth/tumor remission. *Invasion Metastasis,* 16:247–267, 1996.

Adam JA. *Mathematics in Nature: Modeling Patterns in the Natural World.* Princeton, NJ: Princeton University Press, 2003, 2006.

Arnold JA. A simplified model of wound healing: II. The critical size defect in two dimensions. *Math Comput Model*, 30:47–60, 1999.

Chaplain MAJ, Anderson ARA. Mathematical modelling, simulation and prediction of tumour-induced angiogenesis. *Invasion Metastasis*, 16:222–234, 1996.

Foulds L. The experimental study of tumor progression: a review. *Cancer Res*, 14:327–339, 1954.

Greenspan HP. On the self-inhibited growth of cell cultures. *Growth*, 38:81–95, 1974.

Greenspan HP. On the growth and stability of cell cultures and solid tumors. *J Theor Biol*, 56:229–242, 1976.

Greller LD, Tobin FL, Poste G. Tumor heterogeneity and progression: conceptual foundations for modeling. *Invasion Metastasis*, 16:177–208, 1996.

Maggelakis S, Adam JA. Mathematical model for prevascular growth of a spherical carcinoma. *Math Comput Model*, 13:23–38, 1990.

4 Physical Modeling

STACIE I. RINGLEB

INTRODUCTION

In the area of medical and health care modeling and simulation, **physical models** can be defined as tangible physical models of a human being or a piece of the human system, a computational or mathematical model of a physical system within the human body, or a cadaver-based or animal model that is used to help solve a problem in medicine and health care. **Tangible physical models** are the main component of simulators, with varying levels of fidelity, and are in part, task trainers. These simulators are generally used for education and training, in evaluations of new tools, and in assessments of competencies. **Computational physical models** are generally used to answer questions that are difficult to answer experimentally, or they are incorporated into virtual reality simulators used for training and education. Finally, cadaver-based and animal models are used to model physical problems that cannot be tested in living people. **Cadaver-based models** are frequently used to understand how injuries and surgical repair affect the performance of the musculoskeletal system. **Animal-based models** are used to understand topics such as tumor growth and responses to drug treatments and tissue healing under various conditions (e.g., fracture, tendon healing with and without injected compounds).

Modern **physical simulators** of the human being have been used in medical and health care education and training since the 1970s. These models range from full-body medical simulators to specialty-specific simulators, and can be classified in several areas, including physical reality, virtual reality, combinations of physical and virtual reality, and models including haptic feedback to allow users to feel like they are interacting with a person. Additionally, physical models are used to build low-, intermediate-, and high-fidelity simulators. These models of the human system allow students, physicians, and

Modeling and Simulation in the Medical and Health Sciences, First Edition. Edited by John A. Sokolowski and Catherine M. Banks.
© 2011 John Wiley & Sons, Inc. Published 2011 by John Wiley & Sons, Inc.

allied health professionals to train in an environment where they will be exposed to procedures they need to know how to perform. This is especially useful when exposure to certain procedures is limited because no patients are available who require this procedure.

Countless articles have been written on the use of physical simulators in the education and training of medical and health care professionals, and there are many articles assessing the effectiveness of a single simulator. More recently, researchers have been demonstrating the need for more in-depth evaluation of the use of these physical models. Understanding the use of high- vs. low-fidelity simulators and how physical models are being used in situations highlighting the context of their use (e.g., under external stress, such as being observed by a supervisor) are both topics seen in the more recent literature [1].

Computational models of the physical human system have been developed for cells, organs, the musculoskeletal system, blood flow, and so on. Once these models are developed, verified, and validated, they can be incorporated into virtual reality simulators, used to improve our understanding of the behavior of systems within the human body that are difficult to study experimentally, and to understand disease and injuries to the human body. Just as in many other areas of modeling and simulation, there are many ongoing efforts to build useful frameworks for physical models within the human system.

Cadaver models are most frequently used in the area of orthopedic research, where measurements in living humans are not possible because they are too invasive or require the creation of an injury and simulated surgery. Animals are frequently modified to model a human disease state such as cancer, where treatments can be tested before they are safe for humans. Animals are also used in orthopedic applications, such as to create injuries and assess healing under various conditions.

TANGIBLE PHYSICAL MODELS

Physical models are used in all areas for the training and education of physicians and health care professionals to simulate various aspects of patient care, and range from a full-body manikin to specific systems within the body. In the area of medical education, there are fewer patients and instruction time, more highly technical procedures (e.g., laproscopic procedures), and an increased emphasis on patient safety [2], which creates a demand for accurate physical models to be included in simulators used for education and training. Simulators are also used with other health care professionals for education and training, including, but not limited to, nurses, paramedics, and phlebotomists.

A variety of simulators, of low to high fidelity, have been shown to improve performance [3]. However, not all simulators are effective; some have proven to be ineffective [4], and sometimes a low-fidelity simulator is just as good as a high-fidelity simulator [5].

High-fidelity human simulators are used to train physicians [6], nurses [3], and other health care professionals [7–9]. One study reported on their use by faculty members to prepare nursing students sufficiently for work in a clinic [10]. They have also been used to assess treatment protocols [11,12], to assess competencies [13–15], and to improve communication skills [8,16]. When used to evaluate the usefulness of various tools in performing procedures [15,17–20], they have been found to improve skills for at least six to eight weeks after simulator-based training [21]. However, these simulators have to be used with caution, as they are not always completely realistic. For example, one physical model was found not to have a completely realistic physical model of the human airway, which the user must be aware of to maximize the benefit of training with this simulator [22].

Specific Examples of Physical Models Used in Simulators

The two most documented applications in the literature of simulators are high-fidelity advanced patient simulators and surgical simulators. **High-fidelity advanced patient simulators** are complete physical models of the human body. These models are able to combine physiological and pharmological responses. In other words, they allow the user to distinguish heart and lung sounds, reactive eyes, a realistic bronchial tree, the ability to respond to the administration of drugs, and realistic airways for practicing intubations in a variety of situations (e.g., locked neck, difficult airway intubations, clear airway intubations). The simulators allow for procedures including chest tube insertion and cricothyrotomy [i.e., an emergency incision to secure the airway (this is not a tracheotomy)]. Additionally, the software included with these simulators allows scenarios to be programmed in and the performance of the user to be assessed.

A significant amount of research was performed using the airway in these advanced patient simulators. The airway of one simulator, SimMan (Laerdal Medical, Stavanger, Norway), was evaluated and found to be generally acceptable; however, there were some differences, including the fact that the distance from the teeth to the vallecula was too short and some procedures were more different to perform than they are on a living person [22]. Knowing these differences between the simulator and real life will allow the user to obtain full benefits from simulators [22]. The main uses of the airway of an advanced patient simulator are to train clinicians [17,23], to assess the competencies of clinicians [24], and to evaluate different tools for intubation [15,18,19,25,26]. The groups of clinicians most often reported to have been trained and assessed using an advanced patient simulator are medical students and residents, paramedics, nurse anesthetists, and respiratory therapists.

There are countless **surgical simulators** on the market, many of which focus on laparoscopic procedures. Simulators are necessary in order to gain proficiency in many procedures. In the case of laparoscopic surgery, surgeons have limited depth perception, a limited field of view, and difficulty with hand–eye coordination because they are working with instruments that are performing the procedure [27]. Each simulator incorporates its own model of a human system. Some of the physical models within these simulators are validated, whereas others are not; therefore, it is critical to obtain information about the model contained within the simulator and determine if independent assessment studies have been completed on the product before committing to its use [28]. Examples of physical models in surgical simulators include cholecystectomy [29], bronchoscopy [30], lobectomy [31], endoscopy [32, 33], and cardiac catheterization [34].

Surgical simulators may be based solely on virtual reality or augmented reality. **Augmented reality** is a combination of virtual reality and reality, where the realism of the system may come from haptic feedback. In some cases, simulators that include haptic feedback are significantly better than virtual reality simulators [35], and systems with haptic feedback have been shown to improve performance compared to the performance of people not using these simulators [36]. However, many simulators that include haptic feedback require improvements before they can be classified as realistic simulators [32,37,38].

Toward the development of valid simulators with haptic feedback, the material properties of the tissues with which the physician interacts must be measured and modeled mathematically. It has been asserted that surgical simulators without correct soft tissue mechanics are "glorified video games with limited training value" [27]. In other words, it is necessary to have a valid physical model within a simulator or it may have limited or no value for training clinicians effectively. Human tissue has complicated material properties; thus, significant effort has been put into characterizing within our bodies the mechanical properties of the nonlinear, inhomogeneous, anisotropic, and rate-dependent materials in the field of biomechanics. Many researchers have spent their careers developing mathematical equations that can describe the complexities of these material properties but are not computationally expensive. The theory from which many researchers build their models is the **quasilinear viscoelastic theory** developed by Y. C. Fung [39]. This theory has been applied to most of the soft tissues in the body and has been incorporated into more complex equations, such as a biphasic conewise linear elastic–quasilinear viscoelastic model used to model the behavior of articular cartilage [40].

There have been efforts to characterize the material properties of organs [27] and the thigh [38], specifically for the purpose of having simulators based on accurate physics-based models of the human body. Linear elastic and quasilinear viscoelastic models of the liver and stomach were determined by measuring the load displacement characteristics of these organs. These measurements were obtained by integrating a six-axis force transducer with a force resolution of 0.78 mN and a maximum force of 8.5 N into a robotic device with a position resolution of 30 μm. The load–displacement data were collected using a ramp-and-hold technique, where the tissue was indented from 1 to 8 mm at velocities ranging from 1 to 8 mm/s, and held for 60 s. Additional data were collected where the tissues were loaded sinusoidally. The steady-state values from the ramp-and-hold method were used to calculate an effective Young's modulus (i.e., a description of the linear elastic behavior of the tissue), which was 5.94 kPa and 1.91 kPa for the liver and stomach, respectively. The experimental data were then processed to determine the quasilinear viscoelastic force–displacement behavior of these tissues, which was characterized as $F^e(\delta) = A(e^{B\delta} - 1)$, where $A_{liver} = 0.0121$ N, $A_{stomach} = 0.0315$ N, $B_{liver} = 0.9809$, and $B_{stomach} = 0.4580$ [27]. The linear elastic properties of the tissue can be used for situations where small tissue displacements are present, whereas the quasilinear viscoelastic behavior is more appropriate when larger deformations would be applied in reality.

The material properties of the thigh were assessed to develop a physical model that can be incorporated into a simulator to perform echography, an ultrasound-based procedure used to detect thrombosis [38]. This simulator is needed because it takes approximately 1000 examinations before a technician is proficient at determining if a thrombosis is present, and it is common for the first 500 examinations to be supervised by an experienced technician [38]. If a valid simulator existed, this would save considerable time and money in training new technicians, as they would not require as many supervised examinations once the new technician is testing patients. The first step toward developing a simulator for echography was to characterize the material properties of the thigh as they would be felt by an ultrasound technician when applying the probe to a thigh. To characterize these material properties, a force transducer was incorporated into a device with the same contact area as a probe used for echography and with a probe of pyramidal shape, thus determining how the load–displacement characteristics of the thigh differ with various types of probes. This will allow for the development of a more robust simulator. The load–displacement data were then used to develop a two-layer model of the thigh. The surface of the thigh model consisted of masses, linear springs, and dampers. Orthogonal to the surface of the model, a set of nonlinear springs were included. Both groups had a significant improvement in their performance of the procedure on a cadaver after training on either the low- or high-fidelity simulator, indicating that the low- and high-fidelity simulators are equally beneficial [40]. Similarly, when respiratory therapists trained on low- and high-fidelity simulators, there was no difference in the time of intubation on humans [23].

Two simulators for phlebotomy were compared [4]. The first was a simulated limb that included tubing in the place of veins and red liquid to simulate blood when the needle was inserted properly. The second simulator was a computer-based haptic simulator. Students were split into two groups, each of which drew blood from the simulator to which they were assigned, (1) with no training and (2) after training on the same simulator. Then each group went into a clinic to perform blood draws on patients. Although both groups improved with training on each simulator, the group that trained on the higher-fidelity, haptic-based simulator performed significantly worse than did the group trained on the physical model of the arm. The reason for this discrepancy was that the computer-based simulation did not have a physical arm and the students going through the training program did not receive the necessary reinforcement required to tie a tourniquet to the arm before drawing blood [4]. This study reinforced the fact that higher-fidelity simulators, which include more mathematically complicated physical models of human tissue, may not be the best way to train a clinician.

Although the aforementioned studies support the fact that a simpler model can be as effective as the high-fidelity model, one must keep in mind that the high-fidelity model also has other capabilities. For example, one can program a scenario, such as anaphylactic shock, where the clinician must administer epinephrine and intravenous fluids as well as performing a surgical procedure on the airway [42], which is not possible on a low-fidelity simulator. An additional benefit of the high-fidelity patient simulator is that the evaluation capabilities are able to differentiate between novice and experienced residents [42].

CADAVER-BASED MODELS

Cadaver-based models are most commonly used to train medical professionals to perform medical procedures [39] and in orthopedic research. In the area of training, a cadaver is simply a surrogate for a living person, as the anatomy and tissue properties are the same, and they can be useful in learning how to perform a complex procedure on a cadaver, especially if a simulator is not available. The cadaver-based models used in orthopedic research are generally used to assess differences in kinematics (i.e., the movements of bones), joint contact pressures, and tendon excursions (displacements) when injury and treatment are modeled. Injuries are typically modeled by cutting or damaging the tissues within the body, but they are also created by offloading tendons to simulate the dysfunction of a tendon. They may include cutting ligaments or fracturing a bone. These injured models are then treated with simulated surgeries (e.g., repairing ligaments, transferring tendons, adding plates to repair bones).

Cadaver models are extremely useful, as one cadaver can be tested in uninjured condition, injured condition, and treated condition, usually in one to two days of testing. This is advantageous over the testing of living humans, as you rarely have data before injury on a patient, and it frequently takes months after surgery before you could collect data on a patient after surgery. Additionally, in a cadaver, you can test multiple treatments in one specimen. A cadaver model is advantageous over a simulator type of model in many cases, especially in orthopedic research, as the anatomy is exact and the tissue properties are the same, as long as the cadaver is tested without embalming. There are, of course, disadvantages to cadaver testing. For example, you cannot measure the effects of tissue healing, and cadavers are usually not capable of withstanding the levels of loading that living people are able to withstand. There are countless studies involving cadaver models in the joints throughout the body, including the spine, shoulder, elbow, wrist, hip, knee, and ankle. Many of the concepts from testing joint to joint are very similar; a few samples are presented in this chapter.

Cadaver-based models are frequently used to assess surgeries, such as differences between an open and an arthroscopic procedure [41], anatomic surgical reconstruction vs. tendon transfers for joint restabilization [42], and assessments of new surgical procedures [43]. Open and arthroscopic procedures were compared in the shoulder joint for rotator cuff repairs. Intact specimens were tested, then a rotator cuff injury was created and repaired using one of the two surgical procedures. The kinematic results found that the arthroscopic repair did not provide as much stability as the open repair, although neither technique provided the same stability as that of the intact joint (although tissue healing may enhance the joint stability), and there was a large decrease in range of motion after both techniques [41]. When anatomical and tendon transfer techniques were compared in the reconstruction of an unstable hindfoot (i.e., ankle and subtalar joint), it was found that the tendon transfer overconstrained the joint in rotation but not in translations, whereas the anatomical procedure resulted in satisfactory restoration of the joint kinematics [42]. The pull-out strength of a fixation device developed to minimize the risk of neurological complications in patients with scoliosis was tested and found to have results similar to those of traditional methods, suggesting that this new technique may provide effective stabilization and no more failures than with the traditional technique [43].

Another common use of cadaver-based models is to obtain measurements that are difficult or impossible to obtain in living plople, in the intact, injured, and/or surgically repaired joint. The work of friction in the posterior tibial tendon was calculated by inserting transducers on the proximal and distal ends of the posterior tibial tendon to measure the frictional force and also to measure the excursion (displacement) of the tendon with a potentiometer as it moved throughout its range of motion in an intact and flatfoot model. It was found that when a flatfoot was created by sectioning the peritalar soft tissues in the foot, there was a significant increase in the work of friction, suggesting that the flatfoot will contribute to the progressive degeneration of this tendon [44]. Tendon moment arms are also frequently measured in cadavers because it is difficult to measure the joint motion and tendon excursion required to determine the moment arms in vivo, even with imaging techniques, because of the three-dimensional motion of many joints, such as the shoulder [45,46], and the fact that some tendons cross two joints and it is necessary to fuse one of the two joints in order to understand the contribution of the moment arm to the motion of one joint [47].

Cadaveric Gait Simulators

Several research groups have developed high-fidelity simulators that allow cadaver lower limbs to walk [48–55]. In the knee, **dynamic gait simulators** have been used to determine how much joint contact pressure changes as a function of the amount of the meniscus (a soft tissue support structure) that is torn [56] and to assess the effects of tendon transfers on the kinematics of the lower extremity [57]. Gait simulators are more common when examining the action of the foot and ankle. The effects of orthotics on bone strain has been assessed with a gait simulator [48]. By using a cadaver model of gait, the researchers were able to perform a measurement that was too invasive for use in vivo, but in conditions similar to that of real life (i.e., physiological loading and motions). Additionally, they concluded that the use of custom orthotics may minimize stress fractures because they minimized the strain on the second metatarsal, a common site for stress fractures [48].

A similar cadaveric gait simulator measured joint pressure in the midfoot in cadaver feet with and without diabetes, and determined that diabetic feet had higher joint contact pressures than those of nondiabetic feet. This suggested that diabetic patients are predisposed to a condition called *Charcot neuropathy*, where the arch collapses, resulting in a permanent foot deformity, because these higher contact pressures can degrade the integrity of the joint, leading to the deformity [49].

Modeling the human gait with a cadaveric simulator can also help us understand the effects that different muscles have on the function of the foot and ankle. The function of the flexor hallucus longus (FHL) was studied by simulating gait in a cadaver with unrestricted FHL motion and when the FHL was held 2, 4, and 6 mm proximal to the normal path of tendon excursion (displacement) to simulate restrictive tenosynovitis of the FHL. As the restriction of the FHL increased, the forces in the FHL tendon, first metatarsal, and first metatarsal phalangeal joint increased progressively and significantly [50].

Cadaver models of injury and treatment can also be compared with cadavric gait simulators. The effects of two procedures to treat foot drop were compared. Foot drop was simulated by not loading the extensor tendons during gait in both the simulated foot-drop condition and after the posterior tibial tendon was transferred through the interosseous membrane (IOM procedure) and after the Bridal procedure. The simulator found that both procedures corrected the foot-drop deformity and that there was no difference between the procedures [58]. The effects of fracture repair on the kinematics of the foot and ankle were assessed in a high-fidelity cadaveric gait simulator, where the stance phase of gait was assessed (i.e., the time when the foot is in contact with the ground). This study found that the simulated ankle fracture, which included some soft tissue injuries, caused external rotation and inversion of the talus to increase significantly, while surgery repaired this motion. Despite these significant changes, the authors asserted that there was remarkable stability of the ankle joint during stance in the injured condition, suggesting that more instability may be present during the swing phase of gait. If this is the case, the foot and ankle may be in a compromised position at heel strike, and this position may not have been modeled properly in this simulator [59].

ANIMAL MODELS

Animal models are used widely in medical research. A few applications of animal models in the medical and health care field include drug testing for most diseases from cancer [60] to diabetes [61], basic cancer research [62], understanding tumor metastasis [63], improving treatments in orthopedic surgery [64], and as input and/or validation for computational models [65,66]. Many books could be written on the topic of animal models. For the purposes of this chapter, one example of how animal models are used to help develop and validate physical computational models of fracture healing in cortical bone. Cortical bone is the hard bone on the outer surface of bone. When cortical bone heals, fracture callus forms. Clinically, it is useful to be able to assess the properties of the callus formation in order to assess the success of healing, especially when considering drug treatment and rehabilitation protocols [66]. Therefore, a finite element model was created using micro-computed-tomographic (μCT) images obtained from fractured rat bones. A closed unilateral fracture was created in male Sabra rats, and then repaired using an intramedullary pin (i.e., a pin inserted in the intermedullary canal in the center of the bone). The rats were euthenized at three or four weeks postfracture, and the fractured limb was excised. μCT images were obtained from the excised limb, and the bones were tested in torsion until failure. The torsional rigidity ($N \cdot m^2$/rad), failure torque ($N \cdot m$), failure angle (radians), and failure energy ($N \cdot m \cdot$rad) were measured. Voxel-based finite element models were created from the μCT images, where the images were thresholded based on their gray scale to identify regions of highly mineralized bone, new bone, and soft tissue. Torsion was applied to the finite element model and the computational torsional rigidity was compared to the experimental torsional rigidity, resulting in a correlation coefficient of $r = 0.69$ ($p < 0.001$). The finite element model results correlated better with the experimental results than callus area, bone mineral density, and area moment of inertia ($r < 0.3$, $p > 0.05$), which are common measures of fracture healing [66]. The authors concluded that this finite element method may be a useful tool in fracture healing studies [66].

COMPUTATIONAL PHYSICAL MODELS

The final types of physical models described in this chapter are computation models of physical systems. Similarly to animal models, there are countless computational models applied to the physical system, many of which are included in the simulators discussed previously. Consistent with many areas within the discipline of modeling and simulation, standardized frameworks are needed for developing physical models that can easily be combined and integrated. Frameworks have been published [67–73], and federal funding has been used to promote the development of open-source computational resources for the development of physical models at the molecular (e.g., protein folding, RNA dynamics, myosin dynamics), cellular (e.g., cardiac tissue engineering, determinants of cell shape), organ (e.g., neuromuscular dynamics, cardiovascular dynamics), and system (e.g., musculoskeletal system) levels [74]. However, there are countless programs and methods used to develop physical models of components of the human body. For example, the Web site of the Systems Biology Markup Language (SBML), which is a community-based effort to find a common intermediate format that enables communication of the most essential aspects of models [75], lists over 180 software systems and links to over 500 models of biological systems.

One advantage of the constant development of modeling and simulation software packages is that they make the development of computational models of physical systems easier for scientists to implement [72], which results in the expansion of modeling and simulation in the medical and health care field. Computational physical models are frequently used to help researchers understand the behavior of a system within the human body. In the remainder of the chapter we describe briefly how physical models presented in a computational format are useful in research in the area of medical and health care modeling and simulation at the cellular, organ, and system levels.

Developing a physical model of neural tissue and simulating the interactions between dividing cells that develop into a global network organization of neural tissue is an ideal application of a physical model because it helps researchers to analyze this global behavior, which is difficult to perform in vivo. A framework called CX3D was developed. This platform allowed for modeling of the behavior of neural tissue as it generates. Specifically, neurons arise through the replication and migration of precursors, and these neurons mature into cells and extend into axons and dendrites. Individual neurons are then discretized into elements with accurate material properties representing the soma and neurite. The extracellular space is also included in the physical model and interacts with the neuron. In addition to the physical description of the neuron within the model, growth functions are encoded into the software to simulate growth of the neuron [72]. A physical model of cancer cell behavior, specifically tumor encapsulation and transcapsular spread, have also been developed. This model included physical forces and cellular interactions and allowed researchers to compare two hypotheses of capsule formation [76].

At the organ level, physical models are generally created to understand the behavior of those systems and how they might respond to treatment. For example, a computational fluid dynamics approach was used to quantify the velocity field, wall shear stress, and pressure distribution in a fusiform basilar artery (i.e., an artery with two branches) with an aneurysm and in a healthy artery. The vascular geometry was determined using contrast-enhanced magnetic resonance images. The physical model of the aneurysm determined that there was higher pressure and increased wall shear stress in that artery. In the patient tested, the artery was hemodynamically limited because of the stenosis in the artery, and the model determined that the patient would not benefit from a procedure that would occlude the artery [77]. However, the model was altered to remove the stenosis, and the simulation determined that occlusion of the artery with the stenosis would reduce the flow impact on the aneurism wall. Because the physical model results allow the physician to know the hemodynamic forces, this may improve treatment decisions for patients [77].

Thermal ablation is a minimally invasive method used to remove cancerous tumors. A physical model was developed of a tumor and blood vessel located close to the tumor to determine how much tissue could be ablated without damaging the blood vessel. In addition to the physical model of the tumor and blood vessel, temperature distribution dynamics were described by combining three-dimensional bioheat transport in tissue with a one-dimensional model of convective-dispersive heat transport in the blood vessel [78]. This model, combining a physical model with a computational model of the heat induced with thermal ablation, can be used for treatment planning, and with further research it could guide the intraoperative application of thermal ablation to a tumor near a large blood vessel that cannot be allowed to be damaged [78].

Trabecular bone does not heal in the same way that cortical bone does. Specifically, there is very little callous formation, but trabecular bone heals through a process called *intramembraneous ossification.* A brief description of this process includes the laying down of woven (immature) bone in the fracture gap. Once the fracture gap is filled with homogeneous isotropic woven bone, it then remodels into anisotropic, inhomogeneous trabecular bone [65]. A physical model of the healing process of trabecular bone was developed using finite element analysis combined with fuzzy logic. The model was capable of simulating the entire healing process of trabecular bone, which was sensitive to the direction and amount of load applied during the healing process [65]. This physical model could be combined into a model of cortical bone healing, and different treatment conditions for severe fractures could be tested to determine what the best treatment for healing might be.

The system within the body that is most frequently modeled is the musculoskeletal system. Forward dynamic and inverse dynamic models are most common when modeling movement. The most common motion to model is gait, but other motions, such as wheelchair propulsion, are also commonly modeled. The inputs of a forward dynamic model are either muscle forces or joint moments, and the outputs are the kinematic motions. Inverse dynamics models have kinematics as the input and muscle forces and joint moments as outputs. Both forward dynamic and inverse dynamic simulations usually require optimization to solve these models. In addition to forward dynamic and inverse dynamic models, finite element models are also commonly used to develop models of joints and/or muscles. Because they are computationally expensive, they are not used to model a large piece of the musculoskeletal system; they usually just examine a joint and its soft tissue support structures or a muscle. The most commonly used model of the musculoskeletal system is a lower extremity model including 43 musculotenonous actuators (i.e., muscle–tendon units) with lines of action based on their anatomical relationship to three-dimensional bone models. Each muscle–tendon unit includes a mathematical description of the isometric force–length curve, and the model contains joints at the hip, knee, ankle, and subtalar and matatarsophalangeal joints, and 13 degrees of freedom. The model was validated by comparing joint moments with experimental isometric joint moments [79].

The aforementioned model can be used to study a variety of applications, ranging from the effects of pathologies on gait to the effects of strength training on gait. In one study this model and its associated equations of motion were used to determine the capacity of the muscles to extend the hip and knee joints and the joint flexions induced by gravity during the single-limb stance phase of gait in subjects with crouch gait as a result of cerebral palsy [80]. The simulation found that the capacities of all the major hip and knee extensors except the hamstrings were reduced markedly in a crouched gait posture. Crouch gait also increased the flexion accelerations induced by gravity at the hip and knee throughout single-limb stance. These results help explain why patients with crouch gait expend more energy than do people without a crouch gait. Further, the increased flexion accelerations may explain why crouch gait is progressive in patients with cerebral palsy [80].

A similar musculoskeletal model of the upper extremity was used to assess variability in surgical technique, as a possible explanation of variable outcomes in a tendon transfer technique used to restore pinch function in patients with spinal cord injury [81]. The biomechanical model of the upper extremity was used to investigate differences in surgical techniques from 10 surgeons (determined by obtaining measurements during surgery and surveying the surgeons about their surgical techniques). The model assessed the active muscle force-generating capacity of the transferred tendon at various elbow, wrist, and hand postures used commonly in activities of daily living. The simulation determined that a "tighter" tendon transfer would produce peak muscle force in a more flexed position than would a "looser" tendon transfer [81]. The results are a good example of how a computational model of a physical system within the human body can help clinicians make decisions about how to treat patients.

One limitation of the two aforementioned physical models of the musculoskeletal system are that the muscle models use simple geometric shapes to characterize the arrangement of muscle fibers and tendon (i.e., they use a lumped-parameter model), which may result in an inaccurate representation of changes in muscle fiber length and the resulting force–length behavior of muscles with complex architecture [82]. Three-dimensional finite element models of the rectus femoris and vastus intermedius (quadriceps muscles) were developed from magnetic resonance images to determine how the complex architectures of these muscles affect fiber excursions. Fiber excursions calculated with the finite element model were compared with fiber excursions predicted using the lumped-parameter model, and the lumped-parameter model overestimated the fiber excursions for both muscles [82]. This study suggested that improvements in muscle architecture can improve the accuracy of computer simulations.

Finite element models of joints are able to calculate joint contact kinetics and stress distributions in soft tissues in the musculoskeletal system that a dynamic musculoskeletal model mentioned earlier could not determine. An example of the use of finite element models of joints include the calculation of joint contact pressures, which cannot be measured in vivo or calculated with a forward or inverse dynamic model, and calculation of ligament and other soft tissue strain, which can help identify situations where a patient may be at risk for injury.

Another recent trend in modeling of the musculoskeletal system is the development of patient-specific models, where the goal is generally to optimize treatment by improving our understanding of the patient's ailment. A protocol was developed for building a subject-specific biomechanical model of the knee, which combined magnetic resonance imaging, motion analysis (including kinematics and kinetics), and a three-dimensional finite element model. This protocol was used to study the role of body weight on the stresses and strains induced in the knee articular cartilages and meniscus during single-leg stance and calculations of the stresses and ligament forces induced during the gait cycle [83].

As computational power increases, and computational models can answer more complex questions, it is important to examine how various components of the system are modeled so that specific research questions can be answered as accurately as possible with computational models. Conversely, it is not always necessary to take the computational time to run a complicated model when the research question does not require it.

CONCLUSION

Many physical models are used in the area of medical and health care modeling and simulation, including, tangible models used in training (e.g., manikin simulators and surgical laproscopic simulators), cadaver-based models, animal models and physical computational models. Each of these physical models makes an important contribution to the discipline of medical and health care modeling and simulation.

KEY TERMS

physical model
tangible physical model
computational physical
    model
cadaver-based model
animal-based model

physical simulator
high-fidelity advanced
    human simulator
surgical simulator

augmented reality
quasilinear viscoelastic
    theory
dynamic gait simulator
thermal ablation

REFERENCES

[1] Yee B, et al. Nontechnical skills in anesthesia crisis management with repeated exposure to simulation-based education. *Anesthesiology*, 103(2):241–248, 2005.

[2] Ziv A, Small SD, Wolpe PR. Patient safety and simulation-based medical education. *Med Teach*, 22(5):489–495, 2000.

[3] Alinier G, Hunt WB, Gordon R. Determining the value of simulation in nurse education: study design and initial results. *Nurse Educ Pract*, 4(3):200–207, 2004.

[4] Scerbo MW, et al. The efficacy of a medical virtual reality simulator for training phlebotomy. *Hum Factors*, 48(1):72–84, 2006.

[5] de Giovanni D, Roberts T, Norman G. Relative effectiveness of high- versus low-fidelity simulation in learning heart sounds. *Med Educ*, 43(7):661–668, 2009.

[6] Gordon JA, Oriol NE, Cooper JB. Bringing good teaching cases "to life": a simulator-based medical education service. *Acad Med*, 79(1):23–27, 2004.

[7] Grenvik A, et al. New aspects on critical care medicine training. *Curr Opin Crit Care*, 10(4):233–237, 2004.

[8] MacDowall J. The assessment and treatment of the acutely ill patient: the role of the patient simulator as a teaching tool in the undergraduate programme. *Med Teach*, 28(4):326–329, 2006.

[9] Seybert AL, Barton CM. Simulation-based learning to teach blood pressure assessment to doctor of pharmacy students. *Am J Pharm Educ*, 71(3):48, 2007.

[10] Feingold CE, Calaluce M, Kallen MA. Computerized patient model and simulated clinical experiences: evaluation with baccalaureate nursing students. *J Nurs Educ*, 43(4):156–163, 2004.

[11] Bayley R, et al. Impact of ambulance crew configuration on simulated cardiac arrest resuscitation. *Prehosp Emerg Care*, 12(1):62–68, 2008.

[12] Paskins Z, et al. Design, validation and dissemination of an undergraduate assessment tool using SimMan in simulated medical emergencies. *Med Teach*, 32(1):e12–e17, 2010.

[13] Brett-Fleegler MB, et al. A simulator-based tool that assesses pediatric resident resuscitation competency. *Pediatrics*, 121(3):e597–e603, 2008.

[14] Shilkofski NA, Nelson KL, Hunt EA. Recognition and treatment of unstable supraventricular tachycardia by pediatric residents in a simulation scenario. *Simul Healthcare*, 3(1):4–9, 2008.

[15] Tumpach EA, et al. The King LT versus the Combitube: flight crew performance and preference. *Prehosp Emerg Care*, 13(3):324–328, 2009.

[16] Sleeper JA, Thompson C. The use of hi fidelity simulation to enhance nursing students' therapeutic communication skills. *Int J Nurs Educ Scholarship*, 5: Article 42, 2008.

[17] Aziz M, et al. Video laryngoscopy with the Macintosh video laryngoscope in simulated prehospital scenarios by paramedic students. *Prehosp Emerg Care*, 13(2):251–255, 2009.

[18] Liu L, et al. Tracheal intubation of a difficult airway using Airway Scope, Airtraq, and Macintosh laryngoscope: a comparative manikin study of inexperienced personnel. *Anesth Analg*, 110(4):1049–1055, 2010.

[19] Malik MA, et al. A comparison of the Glidescope, Pentax AWS, and Macintosh laryngoscopes when used by novice personnel: a manikin study. *Can J Anaesth*, 56(11):802–811, 2009.

[20] McElwain J, et al. Determination of the optimal stylet strategy for the C-MAC videolaryngoscope. *Anaesthesia*, 65(4):369–378, 2010.

[21] Kuduvalli PM, et al. Unanticipated difficult airway management in anaesthetised patients: a prospective study of the effect of mannequin training on management strategies and skill retention. *Anaesthesia*, 63(4):364–369, 2008.

[22] Hesselfeldt R, Kristensen MS, Rasmussen LS. Evaluation of the airway of the SimMan full-scale patient simulator. *Acta Anaesthesiol Scand*, 49(9):1339–1345, 2005.

[23] Crabtree NA, et al. Fibreoptic airway training: correlation of simulator performance and clinical skill. *Can J Anaesth*, 55(2):100–104, 2008.

[24] Hunt EA, et al. Delays and errors in cardiopulmonary resuscitation and defibrillation by pediatric residents during simulated cardiopulmonary arrests. *Resuscitation*, 80(7):819–825, 2009.

[25] Chen PT, et al. Instructor-based real-time multimedia medical simulation to update concepts of difficult airway management for experienced airway practitioners. *J Chin Med Assoc*, 71(4):174–179, 2008.

[26] Lim TJ, Lim Y, Liu EH. Evaluation of ease of intubation with the GlideScope or Macintosh laryngoscope by anaesthetists in simulated easy and difficult laryngoscopy. *Anaesthesia*, 60(2):180–183, 2005.

[27] Lim YJ, et al. In situ measurement and modeling of biomechanical response of human cadaveric soft tissues for physics-based surgical simulation. *Surg Endosc*, 23(6):1298–1307, 2009.

[28] Issenberg SB, et al. Simulation and new learning technologies. *Med Teach*, 23(1):16–23, 2001.

[29] Aggarwal R, et al. Development of a virtual reality training curriculum for laparoscopic cholecystectomy. *Br J Surg*, 96(9):1086–1093, 2009.

[30] Blum MG, Powers TW, Sundaresan S. Bronchoscopy simulator effectively prepares junior residents to competently perform basic clinical bronchoscopy. *Ann Thorac Surg*, 78(1):287–291, 2004.

[31] Carter YM, Marshall MB. Open lobectomy simulator is an effective tool for teaching thoracic surgical skills. *Ann Thorac Surg*, 87(5): 1546–1550, 2009; discussion 1551.

[32] Edmond CV, Jr, et al. ENT endoscopic surgical training simulator. *Stud Health Technol Inf*, 39:518–528, 1997.

[33] Fried GM, et al. Proving the value of simulation in laparoscopic surgery. *Ann Surg*, 240(3):518–525, 2004; discussion 525–528.

[34] Gallagher AG, Cates CU. Virtual reality training for the operating room and cardiac catheterisation laboratory. *Lancet*, 364(9444):1538–1540, 2004.

[35] Botden SM, et al. Augmented versus virtual reality laparoscopic simulation: What is the difference? A comparison of the ProMIS augmented reality laparoscopic simulator versus LapSim virtual reality laparoscopic simulator. *World J Surg*, 31(4):764–772, 2007.

[36] Lemole M, et al. Virtual ventriculostomy with 'shifted ventricle': neurosurgery resident surgical skill assessment using a high-fidelity haptic/graphic virtual reality simulator. *Neurol Res*, 31(4):430–431, 2009.

[37] Burdea G, et al. Virtual reality-based training for the diagnosis of prostate cancer. *IEEE Trans Biomed Eng*, 46(10):1253–1260, 1999.

[38] d'Aulignac D, et al. Towards a realistic echographic simulator. *Med Image Anal*, 10(1):71–81, 2006.

[39] Friedman Z, et al. Teaching lifesaving procedures: the impact of model fidelity on acquisition and transfer of cricothyrotomy skills to performance on cadavers. *Anesth Analg*, 107(5):1663–1669, 2008.

[40] Girzadas DV, Jr, et al. High fidelity simulation can discriminate between novice and experienced residents when assessing competency in patient care. *Med Teach*, 29(5):472–476, 2007.

[41] Provencher MT, et al. Arthroscopic versus open rotator interval closure: biomechanical evaluation of stability and motion. *Arthroscopy*, 23(6):583–592, 2007.

[42] Ringleb SI, et al. The effect of ankle ligament damage and surgical reconstructions on the mechanics of the ankle and subtalar joints revealed by three-dimensional stress MRI. *J Orthop Res*, 23(4):743–749, 2005.

[43] Hongo M, et al. Biomechanical evaluation of a new fixation device for the thoracic spine. *Eur Spine J*, 18(8):1213–1219, 2009.

[44] Arai K, et al. The effect of flatfoot deformity and tendon loading on the work of friction measured in the posterior tibial tendon. *Clin Biomech (Bristol, Avon)*, 22(5):592–598, 2007.

[45] Hughes RE, et al. Comparison of two methods for computing abduction moment arms of the rotator cuff. *J Biomech*, 31(2):157–160, 1998.

[46] Kuechle DK, et al. Shoulder muscle moment arms during horizontal flexion and elevation. *J Shoulder Elbow Surg*, 6(5):429–439, 1997.

[47] Piazza SJ, et al. Effects of tensioning errors in split transfers of tibialis anterior and posterior tendons. *J Bone Joint Surg Am*, 85A(5):858–865, 2003.

[48] Meardon SA, et al. Effects of custom and semi-custom foot orthotics on second metatarsal bone strain during dynamic gait simulation. *Foot Ankle Int*, 30(10):998–1004, 2009.

[49] Lee DG, Davis BL. Assessment of the effects of diabetes on midfoot joint pressures using a robotic gait simulator. *Foot Ankle Int*, 30(8):767–772, 2009.

[50] Kirane YM, Michelson JD, Sharkey NA. Contribution of the flexor hallucis longus to loading of the first metatarsal and first metatarsophalangeal joint. *Foot Ankle Int*, 29(4):367–377, 2008.

[51] Iaquinto J, Adelaar RS, Wayne JS. Simulation of contact gait in the cadaveric lower extremity using a novel below knee simulator. *Foot Ankle Int*, 29(1):66–71, 2008.

[52] Werner FW, et al. The effect of valgus/varus malalignment on load distribution in total knee replacements. *J Biomech*, 38(2):349–355, 2005.

[53] Milgrom C, et al. A comparison of bone strain measurements at anatomically relevant sites using surface gauges versus strain gauged bone staples. *J Biomech*, 37(6):947–952, 2004.

[54] Hurschler C, Emmerich J, Wulker N. In vitro simulation of stance phase gait: I. Model verification. *Foot Ankle Int*, 24(8):614–622, 2003.

[55] Aubin PM, Cowley MS, Ledoux WR. Gait simulation via a 6-DOF parallel robot with iterative learning control. *IEEE Trans Biomed Eng*, 55(3):1237–1240, 2008.

[56] Bedi A, et al. Dynamic contact mechanics of the medial meniscus as a function of radial tear, repair, and partial meniscectomy. *J Bone Joint Surg Am*, 92(6):1398–1408, 2010.

[57] Anderson MC, et al. A cadaver knee simulator to evaluate the biomechanics of rectus femoris transfer. *Gait Posture*, 30(1):87–92, 2009.

[58] Jotoku T, et al. *Mechanical efficacy of tendon transfer operations for foot drop*. In *50th Annual Meeting of the Orthopaedic Research Society*, San Francisco, 2004.

[59] Michelson JD, et al. Kinematic behavior of the ankle following malleolar fracture repair in a high-fidelity cadaver model. *J Bone Joint Surg Am*, 84A(11):2029–2038, 2002.

[60] Lubet RA, et al. Screening agents for preventive efficacy in a bladder cancer model: study design, end points, and gefitinib and naproxen efficacy. *J Urol*, 183(4):1598–1603, 2010.

[61] Shoda L, et al. The Type 1 Diabetes PhysioLab Platform: a validated physiologically based mathematical model of pathogenesis in the non-obese diabetic mouse. *Clin Exp Immunol*, 161(2):250–267, 2010.

[62] Szpirer C, Cancer research in rat models. *Methods Mol Biol*, 597:445–458, 2010.

[63] Pathi SP, et al. A novel 3-D mineralized tumor model to study breast cancer bone metastasis. *PLoS One*, 5(1):e8849, 2010.

[64] Zhao C, et al. Effects of a lubricin-containing compound on the results of flexor tendon repair in a canine model in vivo. *J Bone Joint Surg Am*, 92(6):1453–1461, 2010.

[65] Shefelbine SJ, et al. Trabecular bone fracture healing simulation with finite element analysis and fuzzy logic. *J Biomech*, 38(12):2440–2250, 2005.

[66] Shefelbine SJ, et al. Prediction of fracture callus mechanical properties using micro-CT images and voxel-based finite element analysis. *Bone*, 36(3):480–488, 2005.

[67] Geris L, Sloten JV, Oosterwyck HV. Connecting biology and mechanics in fracture healing: an integrated mathematical modeling framework for the study of nonunions. *Biomech Model Mechanobiol*, 9(6):713–724, 2010.

[68] Ghosh S, et al. A semiparametric modeling framework for potential biomarker discovery and the development of metabonomic profiles. *BMC Bioinf*, 9:38, 2008.

[69] Iqbal K, Roy A. A novel theoretical framework for the dynamic stability analysis, movement control, and trajectory generation in a multisegment biomechanical model. *J Biomech Eng*, 131(1):011002, 2009.

[70] Moraru II, et al. Virtual cell modelling and simulation software environment. *IET Syst Biol*, 2(5):352–362, 2008.

[71] Shi H, Fahmi R, Farag AA. Validation framework of the finite element modeling of liver tissue. *Med Image Comput Comput Assist Intervent*, 8(Pt. 1):531–538, 2005.

[72] Zubler F, Douglas R. A framework for modeling the growth and development of neurons and networks. *Front Comput Neurosci*, 3:25, 2009.

[73] Chao EY. Graphic-based musculoskeletal model for biomechanical analyses and animation. *Med Eng Phys*, 25(3):201–212, 2003.

[74] Physics-based Simulation of Biological Structures (Simbios). NIH Center for Biomedical Computing at Stanford University. http://simbios.stanford.edu/.

[75] Systems Biology Mark Up Language. http://sbml.org.

[76] Jackson TL, Byrne MH. A mechanical model of tumor encapsulation and transcapsular spread. *Math Biosci*, 180:307–328, 2002.

[77] Jou LD, et al. Computational approach to quantifying hemodynamic forces in giant cerebral aneurysms. *Am J Neuroradiol*, 24(9):1804–1810, 2003.

[78] Chen X, Saidel GM. Mathematical modeling of thermal ablation in tissue surrounding a large vessel. *J of Biomech Eng*, 131(1):011001–011005, 2009.

[79] Delp SL, et al. An interactive graphics-based model of the lower extremity to study orthopaedic surgical procedures. *IEEE Trans Biomed Eng*, 37(8):757–767, 1990.

[80] Hicks JL, et al. Crouched postures reduce the capacity of muscles to extend the hip and knee during the single-limb stance phase of gait. *J Biomech*, 41(5):960–967, 2008.

[81] Murray WM, et al. Variability in surgical technique for brachioradialis tendon transfer. Evidence and implications. *J Bone Joint Surg Am*, 88(9):2009–2016, 2006.

[82] Blemker SS, Delp SL. Rectus femoris and vastus intermedius fiber excursions predicted by three-dimensional muscle models. *J Biomech*, 39(8):1383–1391, 2006.

[83] Yang NH, et al. Protocol for constructing subject-specific biomechanical models of knee joint. *Comput Methods Biomech Biomed Eng*, 13(5):589–603, 2010.

PART THREE
Modeling and Simulation Applications

5 Humans as Models

C. DONALD COMBS

INTRODUCTION

This book covers models and simulations in medicine and the health sciences. A *model* is defined as a representation of something real—in this case a human being. A *simulation* is a process that provides a method of observing a model in action over time, in this case a human being living his or her life. Individual humans have served as medical models of other humans for more than 2000 years.

The concept of human models is particularly important in the medical and health sciences, based as they are on an **apprentice system of education**. In this system, professionals in training are taught the fundamental knowledge of their profession—facts, theories, evidentiary models—and then the techniques used in the provision of health care. This technical training is conducted on real patients needing care. Thus, one of the fundamental challenges of education in the medical and health professions is the reduction of risk to real patients that is posed by the involvement of learners in the provision of care. Addressing this challenge is where the use of humans as models becomes important. The use of humans as models helps learners develop a more solid, realistic understanding of human anatomy and physiology *before* they encounter real patients with real problems.

In this chapter we provide a historical overview of the use of humans as models and simulators. Four such uses are explored: the human cadaver and early wax models of it, simulated patients, plastination, and human data sets. Each use is explored in terms of its historical development, deployment in medical and heath professions training, and overall utility as a teaching and research tool. Finally, some observations about the future use of humans as models are offered.

Modeling and Simulation in the Medical and Health Sciences, First Edition. Edited by John A. Sokolowski and Catherine M. Banks.
© 2011 John Wiley & Sons, Inc. Published 2011 by John Wiley & Sons, Inc.

CADAVERS AND WAX MODELS

The first models of living humans were developed by dissecting dead humans, **cadavers:** that is, dead humans served as models of living humans (Figures 5.1 and 5.2). Galen was a prominent Greek physician who practiced dissection. His writings on anatomy influenced physicians from A.D. 200 to 1500, and acceptance of his emphasis on blood-letting as a remedy persisted well into the nineteenth century. Although Galen wrote primarily about human anatomy, his dissection and research used pigs and apes, particularly the Barbary ape, and was therefore often erroneous [1,2].

In about 1489, Leonardo da Vinci produced a series of anatomical drawings that were far better than Galen's. Over the next 25 years he dissected about 30 human corpses, until Pope Leo X ordered him to stop. His drawings included studies of bone structures, muscles, internal organs, the brain and even the position of the fetus in the womb. His studies of the heart served as a precursor to Harvey's concept of the circulation of the blood.

A few years later, a young medical student, now known as Vesalius, attended anatomy lectures at the University of Paris. There, lecturers explained human anatomy based on Galen's work of more than 1000 years earlier (da Vinci's work was not then widely known) as assistants pointed to the equivalent details in a dissected corpse. Often, the assistants could not find the human

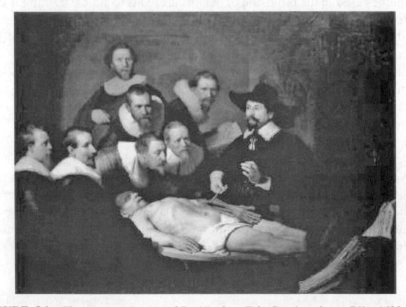

FIGURE 5.1 *The Anatomy Lesson of Dr. Nicolaes Tulp*, Rembrandt van Rijn, 1632. (Reprinted by permission of the Mauritshuis Museum, Mauritshuis, The Hague, The Netherlands.)

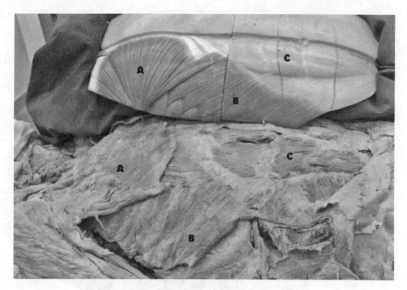

FIGURE 5.2   Muscles of the abdominal wall.

organ as described, but invariably, the lecturers decided that the corpse, rather than Galen, was incorrect! Vesalius decided to dissect human corpses himself and report exactly what he found. Although his approach was controversial, his skill led to appointment as a professor of surgery and anatomy at the University of Padua. In 1540, Vesalius gave a public demonstration of the inaccuracies of Galen's anatomical theories, which remained the orthodoxy of the medical profession. As noted, Galen based much of his analysis on apes. Vesalius, however, compared the skeletons of human beings and apes. He was able to show that in many cases Galen's observations were correct for the apes but not for humans. Vesalius then set about correcting Galen's errors in a series of dissections and drawings. Vesalius published his great work, *De humani corporis fabrica* (The Structure of the Human Body), in 1543 (Figure 5.3). There are seven volumes, including magnificent woodcut illustrations. The book was an immediate success and allowed readers to peer beneath their own skins.

In eighteenth-century Italy, anatomical waxworks were developed to represent the human body. That is, humans served as models for the **wax models** of humans. Life-sized, colored, three-dimensional, soft and moist-looking, they offered compelling replicas of the living body. For those who wanted to see something of anatomy without attending a dissection, the development of these anatomical waxworks offered a realistic alternative. One of the first physicians to use wax models was the French surgeon Guillaume Desnoües (1650–1735), who specialized in educational models. In the early eighteenth

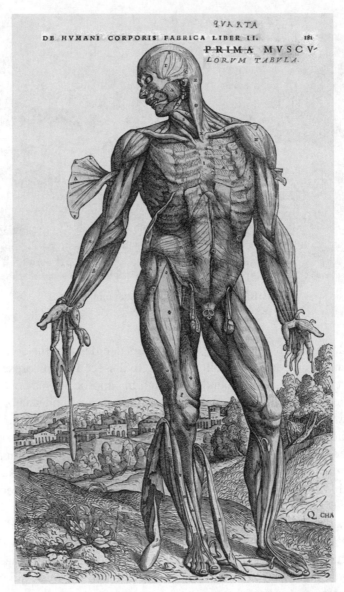

FIGURE 5.3   Vesalius's illustration of the human torso. (Reprinted by permission of the British Library Board.)

century, Ercole Lelli (1702–1766) and his pupil Giovanni Manzolini (1700–1755) founded a school of anatomical wax modeling at the Institute of Sciences of the University of Bologna. A second school of anatomical modeling, founded at the Florentine Museum of Physics and Natural History, later known as "La Specola", eventually contained over 1400 anatomical wax models, including 19 life-sized male and female figures, and was the first museum of its kind open to the public. Wax models did not remove the need for dissection, however. Modeling them was a painstaking process, and hundreds of corpses, which quickly decayed in the Italian heat, had to be dissected to provide the subject matter for just one model [3–6].

These wax models evolved into working models of the human body fashioned in leather and wood. For example, in the mid-eighteenth century, an obstetrical teaching device was fashioned to contain an extraordinary doll-like fetus in utero. These models also simulated the rudiments of childbirth. One model came from the Hospital del Ceppo in Pistoia, near Florence, and others existed in Bologna and Paris. The "baby" could be placed in any position and delivery demonstrated.

The use of dead humans as models for living humans, in the form of either cadavers or wax/ or plastic models, based on analysis of cadavers continues today. It has proved to be an effective tool in helping learners understand human anatomy and the progression of disease. The shortcomings, of course, are grounded in the fact that the models are dead. The models cannot describe what they felt and experienced, and the process of preservation changes the appearance of tissue and organs so that the realism is diminished.

STANDARDIZED PATIENTS

Cadaver based-approaches to modeling the human body were dominate until the 1960s, when more sophisticated approaches—highly trained actors and computer-based manikins—began to appear. Actors trained to mimic patients, called **standardized patients** (SPs), have gone through many refinements since their inception in 1963. Other names have also been used to describe the concept of actors programmed to portray a patient: *patient instructor, patient educator, professional patient, surrogate patient, teaching associate*, and the more generic term *simulated patient*. What all of these terms are referring to is a person who has been carefully trained to take on the characteristics of a real patient, to provide an opportunity for a student to learn or be evaluated on skills firsthand. While working with a standardized patient, the student can experience and practice clinical medicine without jeopardizing the health or welfare of real patients. The value derives from the experience of working with a patient. Using SPs takes learning a step beyond books and a reliance on paper-and-pencil tests. It puts the learning of medicine in the arena of clinical practice—not virtual reality, but veritable reality—as close to the truth of an authentic clinical encounter as one can get without actually being there, because it involves a living, responding human being [7].

The expression *standardized patient*, was coined by the Canadian psychometrician Geoffrey Norman, who was looking for a designation that would capture one of the technique's strongest features, the fact that the patient challenge to each student remains the same. The term was adopted and generally accepted in the 1980s, when the focus of medical education research using simulated patients turned sharply toward research in clinical performance evaluation. The standardized patient offers the student an opportunity to come face to face with the totality of the patient, with his stories, physical symptoms, emotional responses to his illness, attitude toward the medical profession, stresses in coping with life, works and family—in other words, everything a real patient brings to a clinician, overt and hidden (except the necessity of actually "making the patient better"), allowing the student to go about the *process* of unfolding all that she needs to know from the veritable interaction with the patient in order to assist that person to heal.

The standardized patient has today become one of the most pervasive and highly touted of the new methodologies in medical education. A review of the development of SPs provides useful information about the laborious process of using humans as living models and the challenges of ensuring that such use is effective. The following narrative draws heavily on work of Wallace [8,9].

Howard S. Barrows developed the first simulated patient in 1963 when he was teaching third-year neurology clerks at the University of Southern California (USC). Broad acceptance of SPs was slow in coming, however. In fact, during the time that Barrows taught at USC, "no one else was even interested in trying it." Barrows was often invited to speak about neurological subjects, but frequently was requested *not* to talk about simulated patients. Indeed, he was seen as doing something quite detrimental to medical education, maligning its dignity with "actors." When the Associated Press first learned about SPs, the practice was promoted in the popular press with such headlines as: "Hollywood Invades USC Medical School" and descriptions of simulated patients such as: "Scantily clad models are making life a little more interesting for USC medical students." This made it more difficult for Barrows to convince his medical students that the technique was a legitimate educational tool. Resistance persisted even after the 1964 publication of the first article on simulated patients, "The Programmed Patient: A Technique for Appraising Student Performance in Clinical Neurology" in the *Journal of Medical Education.* The USC dean received complaints from medical schools all over the country, but ignored them.

Barrows persisted in using the simulated patient in his clerkship because "it was working." Students loved the technique, and, as he said, "I was learning things about those students I would have never found otherwise." Barrows was searching for an alternative to the traditional method of evaluating students on their clinical clerkships. When faculty got together at the end of a clerkship, Barrows remembers the conversation going something like this: "Let me look at that student's picture . . . Well, I think I remember him." Typically, according to Barrows, most clerks received satisfactory or better evaluations. "And I knew it was because of the way they combed their hair or how neatly they dressed or if they said "Yes sir' and "No sir'." Rarely were student's skills evaluated as unsatisfactory because the faculty almost never directly observed a student with patients. In fact, until the advent of standardized patients, there was no objective clinical metric available with which to evaluate students.

To evaluate the clerks, Barrows needed a case about which he knew everything—all the signs, all the symptoms. He needed a case that could be reproduced for every single clerk in exactly the same way, and he needed someone who had the time and knowledge to record what happened in each encounter with the patient. So he created the first standardized patient case, "Patty Dugger," a paraplegic woman with multiple sclerosis, which was based on a Los Angeles County Hospital patient (Figure 5.4).

After the case was developed, the question of how actually to do the evaluation arose. "Should I peek through a drape, or what should I do? I finally decided that I would make [a] checklist that Rose [the simulated patient] would fill out afterwards." Barrows monitored Rose and the students

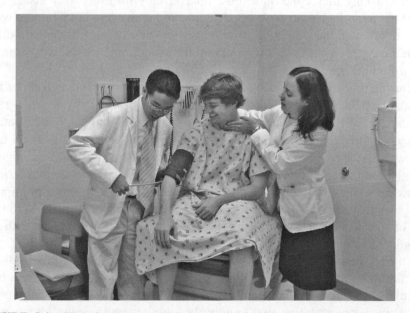

FIGURE 5.4   Clinical encounter utilizing a standardized patient. (Reprinted by permission of the University of Virginia Health System.)

from time to time, but it was Rose who was primarily responsible for recording what happened with each student during every encounter.

Standardized patients arose from the need for a more rigorous method to evaluate the clinical skills of third-year medical students. The methodologies designed by Barrows, from what he called a "pretty primitive" first effort, are the source of the procedures currently being employed by the National Board of Medical Examiners and the Educational Council for Foreign Medical Graduates in their clinical competency exams.

Along with his use of simulated patients to evaluate medical students, Barrows began to see the value of simulated patients in teaching and research. At the same time, he started reaching out to other practicing physicians by designing workshops to help them improve their neurological skills. Barrows' underlying philosophy in these workshops was *experiential learning*, that is, learning by doing and receiving immediate feedback.

Primary among a series of seminal workshops during the 1970s that relied heavily on the use of simulated patients were the "Bedside Clinics in Neurology," sponsored by the American Medical Association. The day before the workshop, Barrows would bring in five prominent neurologists (who would serve as tutors for the workshop participants) and an equal number of simulated patients. The SPs were not only trained to perform several neurological cases, but were also trained to simulate typical continuing medical education participants, such as "the one who isn't interested, the one who's always asking incredible questions, interfering with everybody else." This gave the neurologist-tutors an opportunity to practice and learn how to work effectively with the simulated patients and the neurological cases as well as with the types of physicians with whom they might find themselves working the following day.

During the workshop, each neurologist-tutor was assigned a group of five or six physicians. The challenge was "to make his group of physicians perfect by the end of the day." Each group would start with one simulated patient and work through a case using the time in–time out technique. Time in–time out was first used by Barrows at McMaster University in Canada to enhance small-group teaching. By calling a time out, the interview was put on hold, allowing students an opportunity to discuss among themselves any number of issues that had arisen in the encounter as well as to brainstorm where they might like to take the interview when they went back to time in with the simulated patient.

During the bedside clinics it was the responsibility of the tutors to detect the problems that people in the groups were having and focus their next simulated patient experiences in those areas. As Barrows stated, "If you ask most physicians what they don't know, they don't know they don't know what they don't know."

After seeing a demonstration at the annual meeting of the Society of Directors of Research in Medical Education, Hilliard Jason established a simulated patient program in the first two years of medical school at Michigan State University (MSU). He designed four "difficult patient" cases for the students to experience: a hostile patient, a seductive patient, a patient from another culture, and a patient who hated physicians. During the student interviews, two cameras simultaneously recorded individual shots of the student and the simulated patient. These close-up images were placed electronically side by side in a single, split-screen image so that when Jason later reviewed the videotaped encounter with the student, the actions and the reactions in both the students and the simulated patient could be observed simultaneously. It was one of the first of many educational applications inspired by Barrows' simulated patient work.

As Barrows recalled: "So many faculty teach students to do a complete history and complete physical. There is no such thing. Ask every question, do everything on the physical [exam], there is no such thing. And when they get into real life, they're lucky if they have twenty minutes with a patient. And if they're in an emergency, they're lucky if they've got five minutes. You can't ask every question. . . . So they have to know the right questions to ask."

The discoveries about clinical reasoning were so potent for Barrows that he changed his approach to education, no longer teaching students the "complete" history and physical exam, the way he was taught. Instead, he provided the students with the infinite possibilities a patient provides by letting the students ask anything they wanted, either in direct interaction with a simulated patient or by building that kind of flexibility into written patient problems. In this way, the students learned what questions did and did not have a "payoff" in relation to their hypotheses. The goal was not to memorize an exhaustive list of questions and physical exam maneuvers. It was to guide students into learning what were *appropriate* questions and maneuvers while helping them think through their assumptions of what might be wrong with the patient.

Besides Barrows, there are a number of other threads in the history of standardized patients. Paula Stillman helped to establish the standardized patient as both a credible teaching methodology and a reliable evaluation tool. In the early 1970s, when she was the pediatric clerkship director at the University of Arizona in Tucson, Stillman started using simulated mothers as a technique for teaching interviewing skills. She was inspired by work being done at MSU by another pediatrician, Ray Helfer, who had trained a "programmed mother" to give histories of common pediatric complaints. Helfer, no doubt influenced by the simulated recall research at MSU, employed graduate students to review the tapes of each medical student encounter, then code their behaviors in 25 categories. Stillman found the process complicated and cumbersome. When she returned to Arizona, she was determined to develop a better instrument for teaching and assessing both the content and the process of medical interviewing—an instrument that would be based on behaviors, not abstract ideas, so that it could also be used for giving feedback to students. The Arizona Clinical Interview Rating Scale, or the Arizona Scale as it became known, was the first behaviorally anchored Likert scale to assess medical interviewing skills.

The histories that Stillman taught her simulated mothers to give were compilations of the stories of several children, often including their own, laid out in the format of a checklist. She also taught the mothers to use the checklist to record whether or not a given item was asked, and then to give feedback to the students on their interviewing skills.

"I wasn't doing anything fancy with simulation. It was strictly common pediatric problems and, by that time, well-accepted interviewing skills. The mothers would play the role of the patient and then, at the end, they'd go over the content checklist and the process checklist. In the beginning, I used to videotape everything. But [the mothers] got so good at remembering specifics when they gave feedback that I stopped videotaping.

"By reading the checklists, you couldn't tell what behavior was expected. So I developed a physical exam checklist, with family practitioners and internists, [that] had over 200 items on it. It broke down each component of the physical exam, so when it said 'Examine the Eyes,' there were twenty things you had to do when you examined the eye."

Stillman found a healthy man and a healthy woman, the first "patient instructors," whom she taught to use this checklist. Not only did they know what it felt like when a maneuver was done correctly, but they knew how to teach the student to do it properly. As she explained, "If you weren't reaching up high enough in the axilla when you were palpating the axillary lymph nodes, they could teach you how to do that. They knew nothing about medicine. They were strictly process people." Stillman's patient instructors were not simulating a real patient, they were using their own bodies to teach the medical students how to do a complete, accurate physical examination, using a detailed checklist designed by clinicians.

The only other physician doing anything of similar import was the obstetrician–gynecologist Robert Kretzschmar at the University of Iowa. In 1968, inspired by Barrows' early work with simulated patients, he developed the first *gynecology teaching associates* (GTAs). The GTAs, using their own bodies, were trained to teach students and give them feedback on how to do a proper pelvic exam. In the beginning, the identity of the GTAs, then known as "professional patients," was obscured by covering the women's faces.

The patient's responsibility was to note the various motions and sensations of the physician's examinations and compare each student's performance against these criteria. She therefore gave minimal feedback to the students on their technique. The simulated patient concept, in this rudimentary form, succeeded in providing a conductive environment for instruction with a relaxed, live model, but it did little to enhance communication between student and patient or reliability evaluate a student's technical performance.

By 1972, Kretzschmar had greatly expanded the role of the GTA. No longer were patients' identities masked. The GTA had been given increased responsibilities, including teaching the unique communication skills that go along with the practical skills of a good-quality pelvic examination.

Stillman eventually invited Kretzschmar to the University of Arizona to speak about his work. In the meantime, with her normal physical-exam patient instructors in place in her second-year physical diagnosis course, Stillman knew she "could guarantee that before each student entered his third-year clerkship[s], he could go through a systematic physical exam." She felt confident in the process, until one day she observed a senior medical student who was examining a patient with severe bronchiectasis:

"I said to him, "What do you hear?" And he said, "I don't hear anything. The lungs sound normal." And I said, "Has anybody ever checked out your findings?" And he said, "No, but I listened all over the chest and I percussed." And I realized that I never checked that the students could differentiate normal from abnormal."

This awareness inspired Stillman to search for patient instructors who had actual physical findings: "Tucson had a wonderful population of patients with chronic diseases who were very smart and who had made enough money that they could retire early and really didn't have much to do. I found a man with terrible bronchiectasis who had been an engineer for an astronomical observatory who couldn't work anymore. I found a woman with severe aortic stenosis. I found another one with severe asthma. I found a woman with severe arthritis."

These four were the original patients with chronic findings. Stillman trained these patients to use her normal physical-exam process checklist along with a new content checklist that she designed to take into account the specifics of the actual findings of each patient instructor. She then taught the patients both how to examine themselves and how to teach students to detect the abnormality on their own bodies. For example, in teaching a student, the patient would place the stethoscope properly on her own body until she could hear her own abnormal finding, then she would hold the diaphragm in place while the student listened through his stethoscope. Stillman explained that the patient would then describe in detail the features the student should be listening for: "This is a systolic murmur. First, you [will] hear S1 and then you [will] hear the murmur starting after S1." Stillman thus used humans as models of the effects of disease "because I had this extraordinary cadre of patients. By the time I left Arizona, I must have had seventy-five patients who had chronic stable findings [in] every organ system."

Howard Barrows and Paula Stillman made seminal contributions to the development of standardized patients. Whereas Barrows started using simulation in demonstrations and for summative evaluation, Stillman began her work using patient instructors for teaching and formative assessment. Although both started with checklists, up to the early 1980s, their two approaches to education were different. Barrows' exploration of the principle "learn medicine as you will practice it" led him to incorporate the less tangible elements of the clinical reasoning process into his version of problem-based learning. The veritable encounters he designed for students with simulated patients integrated cognitive learning and practical experience into the "messiness" of human interaction.

Stillman's exploration was based on improving traditional educational methods. She focused on concrete behaviors, thoroughness in the basic skills of interviewing, medical history-taking, and the physical exam to assure that students were prepared for their required clinical rotations. However, for both Barrows and Stillman, the simulated patient became the vehicle by which they were able to investigate their clinical education insights, realize the significant accomplishments of those explorations, and, in so doing, hold the threads until the climate was conductive for others to weave their own investigations.

The 1984 invitational conference, How to Begin Reforming the Medical Curriculum, cosponsored by the Josiah Macy, Jr. Foundation and the Southern Illinois University (SIU) School of Medicine, ignited the adoption of standardized patients, exploding their use in medical schools across the country. Up to that point, simulated patients were seen as not much more than an interesting educational device. This conference provided the impetus to begin scrutinizing the efficacy of evaluating clinical competence by using standardized patient examinations not only as a valuable tool for individual student assessment, but, more potently, as the means for instigating curricular change in medical education.

In an effort to convince deans and associate deans of the usefulness of standardized patients, the Macy Foundation supported a number of follow-up experiential, standardized patient demonstrations. The first of these, Newer Approaches to the Assessment of Clinical Performance, occurred in October 1984, when the attendees of that first invitational conference, which had taken place four months earlier, were invited back to SIU for a hands-on, multiple-station standardized patient demonstration that took place in the first fully equipped dedicated simulated clinic in the country. Designed by Barrows, this *professional development laboratory*, as he called it, became the model for other schools as standardized patient programs grew and the need for dedicated clinic space became a reality.

The major focus of medical education research involving standardized patients from 1984 to the present has been performance assessment. It is the emphasis on performance assessment that has given face validity to the widespread acceptance of this educational innovation.

It is also valuable to clarify the differences between the two types of performance-based assessments: the **objective structured clinical examination** (OSCE) and the **clinical exam** (CPX). The OSCE was introduced in Scotland in the mid-1970s by Ronald Harden of the University of Dundee. The OSCE tests specifically defined skill sets. OSCE station instructions might direct the student to perform a chest exam, take a blood pressure reading on a real patient, take a substance abuse history from a standardized patient, start an IV on a plastic model arm, read an x-ray, or interpret lab results. In Barrows' words, the OSCE assesses the skills of the examinees by "taking a biopsy" of their clinical ability. Station length is usually short (4 to 10 minutes), depending on the complexity of the individual tasks comprising the exam.

On the other hand, the CPX is designed to give students the opportunity to perform with a standardized patient as if they were practicing clinicians in an actual encounter. Students may rely on their own clinical judgment, responding in whatever way seems appropriate based on the patient complaint. The CPX is designed to assess the entire clinical process, including history-taking, appropriately focused physical examination, patient education, and interpersonal skills. CPX stations are generally a minimum of 15 to 20 minutes in length. The cases are portrayed by carefully trained standardized patients. And in a CPX, it is the standardized patient who records the examinee's behavior on a checklist after each encounter. Barrows summarized the differences succinctly: "This [CPX] format focuses on the student's ability to use all the clinical skills to orchestrate them in an appropriate way with appropriate priorities depending upon the problem that was presented. The OSCE can determine whether a student is capable of carrying out a particular skill, but does not determine whether the student will use that skill with an appropriate problem."

The unique demands of this new test modality—using multiple standardized patients who perform the same case within a site or across sites; multiple standardized patient raters using the same checklist; number and length of cases needed for a reliable measure; criteria to determine "clinical competence"; and many more such concerns—challenged the creative thinking of psychometricians, most of whom had previously worked in the more cognitively pure area of multiple-choice examinations.

Simultaneous with these efforts came the National Board of Medical Examiners' Standardized Patient Project, exploring a "high-stakes" clinical assessment component for licensure, the Macy Foundation support of medical school consortia, and the Educational Council for Foreign Medical Graduates' pilot project to assess the clinical competence of graduates of medical schools outside North America who were in U.S. residency programs. All of these efforts were aimed at finding a viable method to accurately assess clinical competence, the consequence of which was the further development, refinement, and ultimate acceptance of standardized patients as that vehicle.

There is an ongoing struggle to integrate the focus of the medical schools—which is to educate its students to the highest standards of excellence—and the concern of the National Board to establish "minimal national standards of clinical competence" for licensure [10–15].

Thus, the use of standardized patients has markedly strengthened the technical preparation of learners before they become directly involved in patient care. It has also allowed the development of performance metrics that supplement teachers' observational assessment of learner competence. Since clinical performance is case-specific, the use of SPs increases the number and variety of cases that learners experience, allowing for a better overall assessment of clinical competence. Standardized patients provide the cases that students need at the time and in the place they are needed. Standardized patient evaluations allow direct comparison of different students' clinical skills. Previously, direct comparisons among students could only be done in the cognitive domain (e.g., through USMLE results). Use of SPs allows students to have a longitudinal experience with patients and to follow a case in a compressed time frame; SPs allow students to be put in clinical situations that they could not manage alone in a real clinical setting; and standardized patient evaluations are responsive to real differences in performance.

As the use of SPs in medical and health professions education has increased, so has the need for research to ensure validity (Does the examination measure what you want to measure?) and reliability (Is a score a reasonable reflection of the examinee's true ability?). Additionally, the need to compensate for the fact that as a general rule, SPs do not have the disease or condition they are modeling has led to the increasingly rapid development of computer-based simulators to complement the capabilities of SPs.

PLASTINATION

A theme throughout the history of medicine has been the quest for knowledge about the interior of the human body. Much of this knowledge has come from cadaver dissection. The knowledge gained has been used in anatomical wax models to preserve structures that in themselves could not be saved.

The use of wax models in the eighteenth century helped overcome two difficulties: the scarcity of cadavers, and the lack of adequate preservation techniques. But waxes, despite their often incredible realism, are a wholly artificial model of the body. Waxes, for the most part, lack the touchability and the immediacy factor. Most are fragile and are meant to be viewed, not handled. Thus, when preservation techniques improved, waxes waned, for preserved bodies could then meet the challenge of cadaver scarcity.

Modern techniques used to preserve the human body for didactic purposes build on methods that began in the time of the Egyptian pharaohs. **Desiccation** was the primary means of mummification and this same general technique was one of the first means used to preserve specimens for teaching. These desiccated specimens included both individual organs and whole-body sets of nerves and vessels. Later improvements came by placing the specimen in alcohol or spirits of wine. Embalming solutions were later developed that were better able to preserve whole bodies. One of the most important of these that greatly improved the quality of teaching specimens was formaldehyde.

A further advancement in preservation of the body was made through replacement of the remaining fluids in an embalmed body with a polymer. This technique was invented in 1977 by Gunther von Hagens, who called the process **plastination** and acquired numerous patents for the process (Figure 5.5). (The following narrative draws on the various writings by von Hagen and Moore cited below.) This technique again builds on earlier ones. Once the body has been embalmed, dissection is performed to expose the parts chosen for display. The body is then placed in an acetone bath. The acetone is absorbed through diffusion and dissolves fats. After the body has been thoroughly impregnated with acetone, it is placed in a polymer such as silicone.

FIGURE 5.5 Gunter von Hagens, inventor of the plastination process, with one of his plastinated torsos. (Reprinted by permission of the Institute for Plastination.)

The acetone is extracted with a steady vacuum, while the polymer is drawn into the body, replacing the acetone. The body is then posed into its desired position, and depending on the polymer used, the plastic is hardened by gas or by heat. The final plastinate has flexibility and a lifelike color and lacks the smell or toxicity of prior preservation methods [16–24].

Traditionally, medical and health professions students familiarized themselves with the human body through a process of removal. First, they remove the skin from the corpse, then detach muscle from the limbs, and finally, remove the chest and abdominal walls. After removing the organs, the remainder of the body is dissected down to the bones and ligaments.

Plastination does not differ much from traditional anatomy. As an innovative preservation method, it makes it possible to create completely new types of specimens. When the polymers harden, for instance, muscles that would ordinarily be slack can provide support, allowing the body to be displayed in a variety of unusual poses, either in its entirety or in various stages of anatomical dissection. It is even possible to take a body and stretch it, thereby creating gaps that reveal structural relationships that would otherwise remain hidden.

Plastinates are able to convey far more than manmade, three-dimensional models, simply because they have come into being via the natural, individual growth of human bodies. Sometimes, plastinates even communicate more than do untreated anatomical specimens. Transparent slices of tissue, for example, allow observers to trace the course of even the most minute nerves into the depths of the body.

Plastination is carried out in many institutions worldwide and has gained acceptance because of the durability of plastinated specimens, the possibility for direct comparison to CT and MR images, and the high teaching value (Figure 5.6).

Plastination has several advantages over other preservation techniques. Compared to desiccation, the plastinated body can preserve the normal appearance: for example, the color, size, and shape of the living being and the important relationships of organs, vessels, and nerves. In comparison to storage in alcohol, the plastinated specimen remains dry and touchable and is more visible. In relation to embalming, plastination removes the factors such as smell and wetness, but the immediacy remains. There is touchability, a sense of authenticity, and, indeed, a certain beauty (Figure 5.6).

Plastinates are valuable in the education of health care professionals. They have been used effectively as adjuncts in the dissecting lab as well as in residency programs. It is possible to preserve excellent dissections, display anatomical abnormalities and disease processes, and demonstrate the results of various surgical procedures. The plastinated specimen has brought dimensionality to teaching in the form of clean, touchable, authentic, nonsmelly, nontoxic, nonbiohazardous specimens.

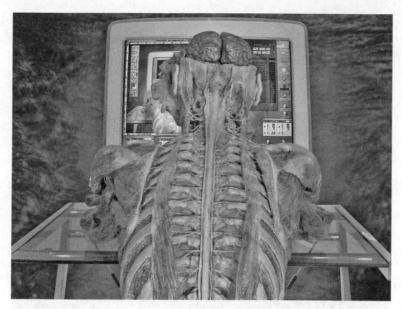

FIGURE 5.6   Plastination virtual web surfer. (Reprinted by permission of the Institute for Plastination.)

HUMAN DATA SETS

Some students of science are beginning to talk in terms of a fourth paradigm for scientific analysis. The first scientific paradigm was empirical and involved increasingly detailed descriptions of natural phenomena. The second paradigm was theoretical, involving the development of models and generalizations based on data observed. The third paradigm begins to harness the power of computers to simulate complex phenomena. The emerging fourth paradigm is data exploration and the discovery of patterns heretofore impossible because of an inability to collect, store, catalog, retrieve, and analyze huge data sets. Data gathering and analysis on a scale not previously possible raises the final concept of humans as models that will be discussed: namely, the human body as a data set.

Today, enormous amounts of data are generated in the provision of health care. Data come from histories and interviews, of course. More to the point, data are also derived from laboratory tests; x-rays; CAT, MRI, and fMRI scans; and ultrasound, just to mention a few sources (Figure 5.7). Increasingly these data are captured in electronic health records on a 24/7 basis and, along with information generated in the artificial world of computer models, are likely to reside forever in a live, substantially publicly accessible, curated state for the purposes of continued analysis.

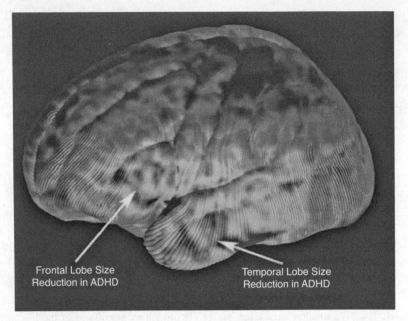

Frontal Lobe Size
Reduction in ADHD

Temporal Lobe Size
Reduction in ADHD

FIGURE 5.7   Three-dimensional high-resolution MRI image of the brain of a patient with ADHD. (Reprinted by permission of Elizabeth Sowell and the UCLA Laboratory of Neuro Imaging.)

There will probably be a time when data will live forever as archival media—just like paper-based storage—and be publicly accessible in the "cloud" to humans and computers. Only recently has a consideration of such permanence for data begun, in the same way that we think of the "stuff" held in our national libraries and museums. These data will not only include individual patient data, but also data from researchers worldwide, past and present.

In 2005, the National Science Board of the National Science Foundation published "Long-Lived Digital Data Collections: Enabling Research and Education in the 21st Century," which began a dialogue about the importance of data preservation and introduced the issue of the care and feeding of an emerging group they identified as "data scientists." "The interests of data scientists—the information and computer scientists, database and software engineers and programmers, disciplinary experts, curators and expert annotators, librarians, archivists, and others, who are crucial to the successful management of a digital data collection—lie in having their creativity and intellectual contributions fully recognized" [25].

Data-intensive science consists of three basic activities: capture, curation, and analysis. One example of humans as data sets that model humans is the Physiome Project of the International Union of Physiological Scientists, which is being used to attempt to provide a comprehensive framework for the human body using computational methods that can incorporate data from biochemistry, biophysical, and anatomical information on cells, tissue, and organs to analyze and build an integrated model of human biological function.

The **Physiome Project** is a worldwide public-domain effort to provide a computational framework for understanding human physiology based on the acquisition and management of huge data sets [26–29]. It aims to develop integrative models at all levels of biological organization, from genes to the whole organism via gene regulatory networks, protein pathways, integrative cell function, and tissue and whole organ structure/function relations. Current projects include the development of:

- Ontologies to organize biological knowledge and access to databases
- Markup languages to encode models of biological structure and function in a standard format for sharing between different application programs and for reuse as components of more comprehensive models
- Databases of structure at the cell, tissue, and organ levels
- Software to render computational models of cell function, such as ion channel electrophysiology, cell signaling and metabolic pathways, transport, motility, and the cell cycle, in two-and three-dimentional graphical form
- Software for displaying and interacting with the organ models, which will allow the user to move across all spatial scales

Eventually, these data sets will be comprehensive enough to populate computer simulations involving tactile holograms that will be programmable to portray the full spectrum of disease and its impact on a functioning human body. When this happens, educators will be able to replace cadavers, plastinated specimens, and SPs with computer software organized to provide the full range of visual, auditory, olfactory, and touch sensations that are now only foreshadowed in today's use of humans as models in the education of medical and health professionals.

CONCLUSION

In this chapter we discuss four cases of humans used as models. The historical use of cadavers, wax models, and plastinates has proved to be effective in modeling the structure of the human body. Their obvious limitation is that there is no longer any process, that is, these uses of humans as models cannot capture either the physiological functions of the human body or the social interactions among humans. Standardized patients allow the incorporation of some physiological modeling and a great deal of social and communicative interactions into the concept of humans as models.

The derivation, integration, and utilization of extensive data on both the anatomy and physiology of humans promise to broaden the use of humans as models, and eventually, once the issues of visualization and tactile holography have been addressed, to allow the digital representation of a range of human models heretofore inconceivable.

KEY TERMS

apprentice system of
  education
cadaver
wax model

standardized patient
objective structured
  clinical examination
clinical exam

desiccation
plastination
plastinates
Physiome Project

REFERENCES

[1] History of Anatomy. http://en.wikipedia.org/wiki/History_of_anatomy. Accessed Dec. 21, 2007.

[2] History of Anatomy. http://www.historyworld.net/wrldhis/PlainTextHistories. asp?groupid=46&HistoryID=aa05. Accessed Dec. 21, 2007.

[3] History of Wax Anatomical Models. hppt://medicina.unica.it/cere/mono01_en. htm. Accessed Dec. 21, 2007.

[4] Lucia D. Women, wax, and anatomy in the 'century of things.' *Renaissance Studi*, 21(4):522–550(29), 2007.

[5] Rosito P, et al. Anna Morandi Manzolini (1716–1774), master sculptress of anatomic wax models. *Pediatri Blood Cancer*, 42(4):388–389, 2004.

[6] Bates AW. Anatomical venuses: the aesthetics of anatomical modeling in 18th- and 19th century Europe. In *Proceedings of International Society for the History of Medicine, 40th International Congress on the History of Medicine*, 2006.

[7] Anderson D. Standardized patients play their part *AAMC Reporter*, Dec. 2006.

[8] Wallace P. Following the threads of an Innovation: the history of standardized patients in medical education. *Caduceus Hum J Med Health Sci*, 13(2):5–28, 1997.

[9] Haggerty ME. The improvement of medical instruction. *J Assoc Am Med Coll*, 4:42–58,1929.

[10] Gorter S, Rethans JJ, Scherpbier A, van der Heijde D, Houben H, van der Vleuten C, van der Linden S. Developing case-specific checklists for standardized-patient-based assessments in internal medicine: a review of the literature. *Acad Medi*, 75 (11):1130–1137, Nov. 2000.

[11] Hirschfelder AD. Coordination in teaching of the fundamental and clinical sciences *J Assoc Am Med Coll*, 4:6–12, 1929.

[12] Culver JO, Bowen DJ, Reynolds SE, Pinsky LE, Press N, Burke W. Breast cancer risk communication: assessment of primary care physicians by standardized patients: *J Genet Med*, Oct. 2009.

[13] Ryan CA, Walshe N, Gaffney R, Shanks A, Burgoyne L, Wiskin CM. Using Standardized patients to assess communication skills in medical and nursing students. *BMC Med Educ*, 10:24, 2010.

[14] Schneider CR, Everett AW, Geelhoed E, Kendall PA, Clifford RM. Measuring the assessment and counseling provided with the supply of nonprescription asthma reliever medication: a simulated patient study. *Ann Pharmacother*, Sept. 2009.

[15] Sutnick MR, Carroll JG. Using patient simulators to teach clinical interviewing skills. *J Am Diet Assoc*, 78(6):614–616, June 1981.

[16] von Hagens G. Impregnation of soft biological specimens with thermosetting resins and elastomers. *Anat Rec*, 194(2):247:55, June 1979.

[17] von Hagens G. The current potential of plastination. *Anat Embryol (Berl)*, 175(4):411–421, 1987.

[18] von Hagens G, Whalley A. *Body Worlds: The Anatomical Exhibition of Real Human Bodies*. Heidelberg, Germany: Institut fur Plastination, 2004 (English translation).

[19] Van Dijck J. Body worlds: t*he art of plastinated cadavers. Configuration*, 9:99–126, 2001.

[20] Sakamoto S, Miyake Y, Kanahara K, Kajita H, Ueki H. Chemically reactive plastination with shin-Etsu silicone KE-108: *J Int Soc Plastination* 21:11–16, 2006.

[21] Latorre RM, Garcia-Sanz MP, Moreno M, Hernandez F, Gil F, Lopez O, Ayala MD, Ramirez G, Vazquez JM, Arencibia, A, Henry RW. How useful is plastination in learning anatomy? *J Vet Med Educ*, 34(2):172–176, Jan. 2007.

[22] Lozanoff S, Lozanoff BK, Sora MC, Rosenheimer J, Keep MF, Tregear J, Saland L, Jacobs J, Saiki S, Alverson D. Anatomy and the access grid: exploiting plastinated brain sections for use in distributed medical education. *Anat Rec B New Anat*, 270(1):30–37, Jan. 2003.

[23] Moore CM, Brown CM. Gunther von Hagens and *Body Worlds*, Part 1: *The Anatomist as Prosektor and Proplastiker*. *Anat Rec B New Anat* 276:8–14, 2004.

[24] Moore CM, Brown CM. Gunther von Hagens and *Body Worlds*, Part 2: T*he Anatomist as Priest and Prophet. Anat Rec B New Anat)* 277:14–20, 2004.

[25] Hey T, Tansley S, Tolle K. The Fourth Paradigm: Data-Intense Scientific Discovery. Redmond, WA,: Microsoft Research, 2009.

[26] Kusumoto L, Heinrichs WL, Dev P, Youngblood P. Avatars alive! The integration of physiology models and computer generated avatars in a multiplayer online simulation. *Stud Health Technol Inf*, 125:256–258, 2007.

[27] Malhotra AG, Darvill E, Pryce-Roberts A, Lundberg K, Konradsen S, Hafstad H. Mind the gap: learner's perspectives on what they learn in communication compared to how they and others behave in the real world. *Patient Educ Couns*, 76(3):385–390, Sept. 2009.

[28] IUPS Physiome Project. http://nwv.physiome.org.nz. Accessed June 1, 2010.

[29] Sowell ER, Thompson PM, Welcome SE, Henkenius AL, Toga AW, Peterson BS. Cortical abnormalities in children and adolescents with attention-deficit hyperactivity disorder. *Lancet*, 362(9397):1699–1707, Nov 2003.

6 Modeling the Human System

MOHAMMED FERDJALLAH and GYUTAE KIM

INTRODUCTION

In human physiology, common organ systems include, among others, the circulatory system, the respiratory system, the immune system, and the nervous system [1]. An **organ system** is a group of organs that are joined together in a functional unit to perform a certain task, whereas an *organ* is a collection of tissues and cells grouped in a structural unit to fulfill a common function. For example, the heart consists of the myocardium, nerves, blood, and connective tissues. Organ systems often share organs which are functionally related and cooperate to form different organ systems. Thus, the organ systems often significantly overlap in their functional aspects. Both nervous and endocrine systems operate via a shared organ, the hypothalamus. The two systems are combined and studied as the neuroendocrine system. Similarly, the musculoskeletal system combines functions of the muscular and skeletal system.

All human societies have medicinal beliefs that attempt to explain and cure diseases. For example, the use of plants as healing agents is a common ancient practice [2]. These medicinal traditions were passed between generations. Ancient Egyptian medicine developed into a practical use in the fields of anatomy, public health, and clinical diagnostics [2]. The earliest known surgery was performed in ancient Egypt [2]. Ancient records suggest that diabetes mellitus was known in early civilizations. Diabetes mellitus is a condition in which a person has a high blood glucose level. Serious long-term complications include cardiovascular disease, chronic renal failure, and retinal damage. It is not surprising that diabetes may have been one of the most common illness in ancient times. Diabetes mellitus appears to have been fatal. However, ancient scientists attempted to treat it, but could not give a good prognosis. In medieval times, scientists described the chronic symptoms of

Modeling and Simulation in the Medical and Health Sciences, First Edition. Edited by John A. Sokolowski and Catherine M. Banks.

diabetes mellitus and documented the sweet taste of urine, and prescribed herbal medicines that are still used in modern medicine. Although diabetes has been recognized since antiquity, and treatments of various efficacies have been known in various regions since the Middle Ages, the pathogenesis of diabetes has been understood experimentally only since the nineteenth century [3]. Early understanding of the human body dominated and influenced medical science for centuries. Early medical anatomy was based on animal models, since human dissection was not permitted. In the nineteenthth century, medicine was improved dramatically by advances in biology, physiology, chemistry, and laboratory techniques [2,4]. The objective of this chapter is to provide a generic framework of organ system modeling and its uses to predict clinical observations and effects of chronic conditions.

ORGAN MODELING

Modeling of **biophysical phenomena** is the ability to create models that validate clinical observations. These models can be used to investigate a variety of medical complications and chronic illnesses. The ultimate use of biological model is to predict and visualize clinical manifestations that may be impossible to reproduce even through clinical assays. The evaluation of whether or not a given mathematical model describes a correct biological system depends on whether the model fits clinical measurements. When a model is designed to match the property of biophysical plan, the predictions of the model could reveal insights that would take longer for experimental knowledge to reveal. In particular, the **Hodgkin and Huxley model** has revealed not only the dynamics of the opening and closure of ion channels, but also provided insights into the shape and conformational changes of the channels long before protein modeling became an advanced science.

For decades, the **immune system** has been the focus of extensive research to design descriptive and predictive models. The human body is constantly being exposed to infectious agents, and yet in most cases the human body is able to resist these infections. The hormonal immune system is able to recognize most foreign patterns using a huge diversity of lymphocyte, a type of white blood cell, with relatively low affinity thresholds. The hormonal immune system incorporates mechanisms that enable lymphocytes to learn the structures of specific foreign proteins. The hormonal immune system evolves and reproduces lymphocytes that have high affinities for specific pathogens. The earliest model proposed to explain this process was that of Kesmir and De Boer [5]. The model attempted to describe the kinetics of lymphocyte's affinity adaptation. However, Kesmir and

De Boer's model has many known parameters that need to be guessed. In addition, the nonlinearity and the large number of equations prohibit an analytical study, which could provide further insights into the behavior of the system.

Another example of organ modeling is hypertension [6,7]. **Hypertension** or high blood pressure is a chronic medical condition in which the blood pressure in the arteries is elevated. Primary hypertension refers to high blood pressure for which no medical cause can be found. Secondary hypertension is caused by conditions that affect the kidneys, arteries, heart, or endocrine system. Uncontrolled high blood pressure increases the risk of heart disease and stroke. High blood pressure typically develops over many years. Hypertension often is part of complications such as diabetes mellitus, combined hyperlipidemia, and central obesity. Thus, the detection of childhood hypertension becomes important in reducing long-term health risks [8]. Developing models that can aid to better understand these kinds of physiological and pathological processes may lead to a significant reduction in associated costs. These models could enhance medical interventions, diagnosis, planning, and treatment procedures.

MODELING OF MUSCULAR ELECTROPHYSIOLOGY

Hodgkin and Huxley Neuron Modeling

A successful model of physiological phenomena was that of Hodgkin and Huxley, which provided a scientific understanding of ion channel dynamics [9–12]. Hodgkin and Huxley designed instrumentation and mathematical models that were able to identify the dynamics of ions channels that participated in the generation and propagation of action potentials decades before the discovery of protein kinetics [9]. The Hodgkin and Huxley model is a set of nonlinear ordinary differential equations that approximates the electrical characteristics of excitable cells such as neurons and myocytes. The components of a typical Hodgkin and Huxley model are shown in Figure 6.1 [9].

By using the Nernst equation, the potential for single ions can be calculated. The goldman equation, an expanded form of the Nernst equation, provides a quantitative expression of the membrane potential for multiple ions. The membrane potential is the voltage difference between the inside and the outside of a cell membrane. Sodium (Na^+), chloride (Cl^-), and potassium (K^+) are the major ions that affect the membrane potential. Based on the Goldman equation, the membrane potential can be expressed by equation (6.3). Equations (6.1) and (6.2) give the general expressions of the Nernst and

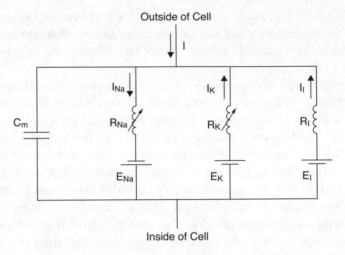

FIGURE 6.1   Hodgkin and Huxley membrane model.

Goldman equations:

$$E_X = \frac{RT}{zF} \ln \frac{[X^+]_{out}}{[X^+]_{in}} \tag{6.1}$$

$$E_{X,Y,Z} = \frac{RT}{F} \ln \frac{P_X[X^+]_{out} + P_Y[Y^+]_{out} + P_Z[Z^-]_{in}}{P_X[X^+]_{in} + P_Y[Y^+]_{in} + P_Z[Z^-]_{out}} \tag{6.2}$$

$$E_{MP} = \frac{RT}{F} \ln \frac{P_K[K^+]_{out} + P_{Na}[Na^+]_{out} + P_{Cl}[Cl^-]_{in}}{P_K[K^+]_{in} + P_{Na}[Na^+]_{in} + P_{Cl}[Cl^-]_{out}} \tag{6.3}$$

where R, T, and F are the gas constant, the absolute temperature in kelvin, and the Faraday constant, respectively. The term z in equation (6.1) is the charge of ion X, and P_X, P_Y, and P_Z are the permeability constants for each ion. $[X^+]_{out}$ and $[X^+]_{in}$ are the concentrations for an ion X outside and inside the cell.

An action potential is a short-lasting event in which the electrical membrane potential of a cell rises and falls rapidly. Action potentials occur in several types of excitable cells, including neurons, muscle cells, and endocrine cells. The ionic currents of the action potential flow in response to concentration differences of the ions across the cell membrane. These concentration differences are established by ion pumps, which use cellular energy to pump ions against their concentration gradient. The ion pump most relevant to the action potential is the sodium–potassium pump. Ion pumps influence the action potential only by establishing the relative ratio of intracellular and extracellular ion concentrations. The action potential involves mainly the opening and closing of ion channels. Hodgkin and Huxley used voltage clamp electrodes and channel blockers to select and study ionic current separately, and thus they were able to derive mathematical expressions for each ionic current [10–12]. The fundamental equation (6.4) governing the membrane potential is

$$I_{total} = C_m \frac{dV_m}{dt} + g_{Na}(V_m - V_{Na}) + g_K(V_m - V_K) + g_L(V_m - V_L) \qquad (6.4)$$

I_{total} is the total membrane current, C_m the membrane capacity per unit area, V_m the displacement of the membrane potential from its resting value, and V_{Na}, V_K, and V_L are potentials measured independently for Na^+, K^+, and leakage, respectively. Theoretically, these potentials can be calculated by the displacement between the absolute value of the resting potential and equilibrium potential for each ion from the Nernst equation [equation (6.1)]. The ion conductance is expressed as g_{ion}. In the Hodgkin and Huxley model, ion conductance can be expressed by multiplying the maximum conductance $(\overline{g}_{Na}, \overline{g}_K, \overline{g}_L)$ and the probability value, which is between 0 and 1.

Each K^+ channel has four gates, which are identical and operate independently. The channel is open only when all gates are open. The rate constant of opening and closing the gate is a function of the voltage. The conductance of K^+ rises monotonically to a steady level as long as a step voltage is maintained, whereas the conductance of Na^+ rises and drops even if the step voltage is maintained. Hodgkin and Huxley theorized that there are two types of gates, fast (m-gates) and slow (h-gates). The m-gates open when the membrane potential increases and the h-gates close when the membrane potential increases [9,10,12]. Each Na^+ channel has three m-gates and one h-gate. The rate constants and constants were calculated from a curve-fitting formula to the data collected by Hodgkin and Huxley. They used these expressions along with the fundamental equation to regenerate the action potential and thus provided a systematic methodology to verify their models. The conductance for each ion is given as follows:

$$g_{Na} = \overline{g}_{Na}m^3h, \qquad g_K = \overline{g}_K n^4, \qquad g_L = \overline{g}_L$$

where $dm/dt = \alpha_m(1 - m) - \beta_m m$, $dh/dt = \alpha_h(1 - h) - \beta_h h$, and $dn/dt = \alpha_n(1 - n) - \beta_n n$. α_m, α_h, α_n, β_m, β_h, and β_n are the ion transfer rates through the membrane. Hodgkin and Huxley proposed the solutions for n, m, and h based on many assumptions and clinical measurements. To demonstrate the model, Hodgkin and Huxley designed a set of experimental essays and novel instrumentation (i.e., the voltage clamp electrode) to estimate the conductance and regenerate the action potentials using estimation methods.

Electrophysiological Muscle Modeling

The Hodgkin and Huxley model was derived for a squid axon 400 to 800 μm in diameter [9]. However, the muscle fiber is less than 100 μm [13], and thus it would have been a challenge for Hodgkin and Huxley to perform experimental essays on muscle fibers. Although the technology is possible today, estimating the ion conductance using a voltage clamp electrode would still be cumbersome, but not impossible. Therefore, an alternative modeling of muscle action potential is warranted to determine the dynamics of muscle ion channels.

Functionally, muscle fibers are connected in a motor unit. A **motor unit** is composed of one motor neuron and the skeletal muscle fibers that it innervates. Although the components of a motor unit are located in one muscle, they may be scattered throughout the muscle. On a muscle fiber, there is a specialized zone, the neuromuscular junction, which is a communication link between a motor neuron and a muscle fiber. The axon of a motor neuron innervates a skeletal muscle fiber through this site. The flow of ions through the channels initiates the contraction and relaxation of skeletal muscles. Voltage-gated channels open and close based on the changes in the membrane potential. Ligand-gated channels are operated by the attachment of a specific ligand to the channels, as shown in Figure 6.2. In skeletal muscle cells, sodium (Na^+) and potassium (K^+) ions play a main role in maintaining the membrane

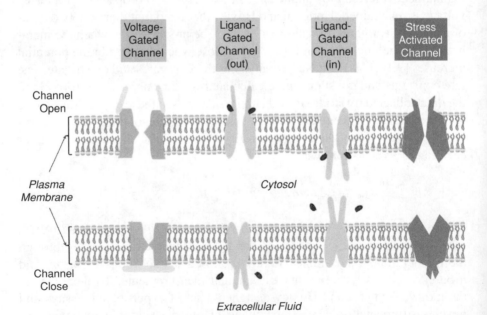

FIGURE 6.2 Various types of ion channels.

potential. Generally, the normal range of the resting membrane potential is between -80 and -100 mV, and the ion potentials can vary depending on the resting membrane potential.

Muscle contraction begins with a transmitted trigger signal, the nerve action potential (AP), from the central nervous system. When a nerve AP reaches the end of a motor neuron, it stimulates acetylcholine vesicles, which contain many acetylcholine (ACh) molecules. The AChs released bind to the receptors on the motor end plate, causing the Na^+ and Cl^- ion channels to open. Changes in the concentrations of ions continuously generate a muscle AP that propagates along the muscle fiber until it becomes extinct at the tendon.

Intracellular Action Potential Modeling

Action potential in muscle is generated by the selective permeability of ion channels. In a resting phase, different concentrations of ions between inside and outside of cell membrane maintain the resting membrane potential. During muscle cell excitation, the membrane is more permeable to Na^+ ions; more Na^+ ions flow into the cell. Consequently, the interior membrane potential increases and triggers the membrane depolarization. At the same time, membrane becomes more permeable to K^+ ions; more K^+ ions flow out of the cell. The inflow of Na^+ ions and the outflow of K^+ ions bring the membrane potential to its resting membrane potential. These ion movements shape the general profile of the muscle action potential. The intracellular action potential (IAP) is generated during cell excitation, as shown in Figure 6.3. IAP is used as a unit source to generate electromyography (EMG). EMG is the study of muscle function through analysis of the electrical signals emanated during voluntary or involuntary muscle contractions. EMG is useful in interpreting pathological states of the musculoskeletal system. IAP is also useful in predicting the transmembrane current, which is the circulating current through the cell membrane. The transmembrane current can be computed mathematically from the second derivative of IAP. Thus, EMG generated from individual IAPs can provide new alternatives to analyze EMG. Computer-generated EMG can be used to model and simulate various cases, such as muscle fatigue, muscle atrophy, motion control, and rehabilitation.

During the last few decades, IAP modeling has been studied extensively to understand the cellular mechanisms of the action potential [14–17]. However, many IAP models require complex computations that hinder their uses for modeling an entire muscle. Due to its mathematical simplicity and computational efficiency, Rosenfalck's model has been widely used [16,17]. Rosenfalck's mathematical expression for muscle IAP is based mainly on data pattern curve fitting [17]. The IAP model by Rosenfalck is given by the

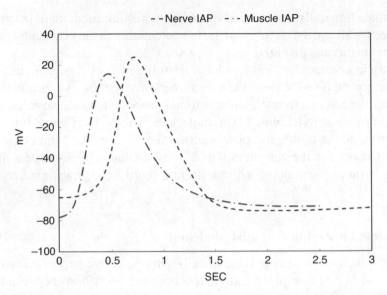

FIGURE 6.3   General intracellular action potential in muscle (dashed-dotted line) and nerve (dashed line).

following empirical expression:

$$V_{IAP}(z) = V_{sc}z^3e^{-\lambda z} - V_{rm}$$

where z is the distance of IAP propagation, V_{sc} a scaling factor based on experimental data, V_{rm} the resting membrane potential, λ a factor representing the rate of repolarization, and V_{sc}, V_{rm}, and λ are adjusted parameters based on experimental data. However, when computed from the Rosenfalck's IAP model, the transmembrane current did not match the experimental data. These descripencies were partially adjusted by Nandedkar IAP model [16]. Consequently, a mathematical model that accurately duplicates the IAP is warranted for EMG generation. The model of three modified gamma distribution functions was designed to represent accurately the IAP in muscle [18]. From this model, bioelectrical signals such as transmembrane current, single muscle fiber action potential, and EMG can be generated to provide insights into the state functions of membrane channels at the cellular level. The general gamma probability density function can be given as

$$G_{mod}(x) = \gamma x^\alpha e^{-\beta x}$$

where α a shape parameter, β a rate parameter, and γ a scaling parameter. Rosenfalck's IAP model uses only one gamma distribution function and thus did not have any biochemical meanings. In contrast, a model with three gamma distribution functions may provide insights into the cellular mechanism that governs the generation and propagation of IAP [18]. The generalized IAP model can be expressed

$$G(x) = \sum_{n=1}^{3} \gamma_n x^{\alpha_n} e^{-\beta_n x} = \gamma_1 x^{\alpha_1} e^{-\beta_1 x} + \gamma_2 x^{\alpha_2} e^{-\beta_2 x} + \gamma_3 x^{\alpha_3} e^{-\beta_3 x}$$

where α_1, α_2, α_3, β_1, β_2, β_3, γ_1, γ_2, and γ_3 are the parameters for the three independent modified gamma distribution functions. To estimate the parameters in the new IAP model, a nonlinear least-squares problem is solved by iterative computation using Newton's method with the Levenberg–Marquardt algorithm. The goal of this algorithm is to search for solution based on the minimum sum-of-squares residue (SSR) method. Figure 6.4 illustrates measured IAPs, which were obtained by digitizing their original signals. These digitized IAPs were set as desired signals for the nonlinear least-squares optimization techniques.

Given the solution vector B, $[\alpha_1, \alpha_2, \alpha_3, \beta_1, \beta_2, \beta_3, \gamma_1, \gamma_2, \gamma_3]$, the adaptive increments in the parameters of the model are computed as

$$\Delta B = \varepsilon \cdot \mathrm{inv}(H + \mu D) \cdot g$$

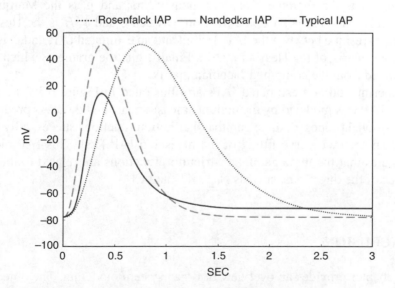

FIGURE 6.4 Digitized muscle IAP (solid line), Rosenfalck IAP model (dotted line), and Nandedkar model (dashed line). Rosenfalck's and Nandedkar's rest membrane potentials were adjusted to that of the digitized IAP. A conduction velocity of 4 m/s was selected.

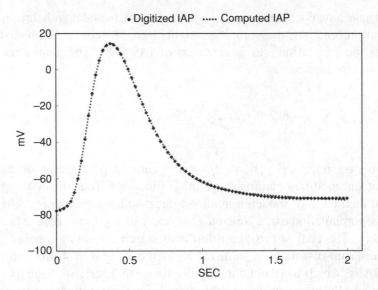

FIGURE 6.5 Digitized muscle IAP (small squares) and curve-fitted muscle IAP (dotted line). The digitized signal was obtained from the EDL muscle of a rat. There was no hyperpolarized period during phase changes. The initial value of the resting membrane potential, was approximately −77.79 mV. The maximum value was 14.44 mV. The curve-fitted IAP was computed by solving a nonlinear least-squares regression.

where ΔB is the increment or decrement vector and μ is the Marquardt parameter. The parameter ε is a fraction between 0 and 1. H is the Hessian matrix of residue between the IAP desired and an estimated IAP, and D is the diagonal matrix of the Hessian matrix. Finally, g is the gradient, which was calculated from the computed Jacobian matrix.

Experimental and estimated IAPs are illustrated in Figure 6.5. The estimated IAP was modeled by the nonlinear least-squares method. As predicted by the model proposed, three gamma distribution functions successfully provided an accurate curve fitting of the measured IAP [18]. This IAP model postulates that the three gamma distribution functions are related to the dynamics of the three ions, such as Na^+, K^+, and Cl^-.

CONCLUSION

This chapter provides an overview of organ system modeling. The objective of organ system modeling is to suggest insights into the concurrent physiological activities that may not be possible through clinical measurements. The Hodgkin and Huxley model is a prime example of biophysical modeling.

Using mathematical equations, a muscle model is presented to investigate the dynamics of ion channels.

KEY TERMS

organ system
biophysical phenomena
Hodgkin and Huxley model

immune system
hypertension
modeling

muscle electrophysiology
motor unit

REFERENCES

[1] Randall D, Burggren W, French, K. *Eckert Animal Physiology: Mechanism and Adaptations*. New York: W.H. Freeman, 2002.

[2] Al-Achi. *An Introduction to Botanical Medicines: History, Science, Uses, and Dangers*. London: Greenwood, 2008.

[3] De Fronzo RA, Ferannini E, Keen H, Zimmet P. *International Textbook of Diabetes Mellitus*. Hoboken, NJ: Wiley, 2004.

[4] Dawson I. *Medicine in the Middle Ages*. London: Hodder Wayland, 2005.

[5] Kesmir C, De Boer RJ. A mathematical model on germinal center kinetics and termination. *Immunol*, 163:2463–2469, 1999.

[6] Badyal DK, Lata H, Dadhich AP. Animal models of hypertensions and effect of drugs. *Indian J Pharmacol*, 35:349–362, 2003.

[7] Ganten D, De Jong W. *Experimental and Genetic Models of Hypertension*. New York: Elsevier, 1994.

[8] Luma GB, Spiotta RP. Hypertension in children and adolescents. *Am Fam Physician*, 73:1558–1568, 2006.

[9] Hodgkin AL, Huxley AF. A quantitative description of membrane current and its application to conduction and excitation in nerve. *J Physiol*, 117:500–544, 1952.

[10] Hodgkin AL, Huxley AF. Current carried by sodium and potassium ions through the membrane of the giant axon loligo. *J Physiol*, 116:449–472, 1952.

[11] Hodgkin AL, Huxley AF. The components of membrane conductance in the giant axon loligo. *J Physiol*, 116:473–496, 1952.

[12] Hodgkin AL, Huxley AF. The dual effect of membrane potential on sodium conductance in the giant axon loligo. *J Physiol*, 116:497–506, 1952.

[13] Frontera WR, Reid KF, Phillips EM, Krivickas LS, Hughes VA, Roubenoff R, Fielding RA. Muscle fiber size and function in elderly humans: a longitudinal study. *J Appl Physiol*, 105:637–642, 2008.

[14] Andreassen S, Rosenfalck A. Relationship of intracellular and extracellular action potentials of skeletal muscle fibers, *Criti Rev Bioeng*, 6:26–36, 1981.

[15] Fleisher SM. Mathematical model of the single-fiber action potential. *Med Biol Eng Comput*, 22:433–439, 1984.

[16] Nandedkar SD, Stalberg E. Simulation of single muscle fiber action potentials. *Med Biol Engi Comput*, 21:158–165, 1983.

[17] Rosenfalck P. Intra- and extracellular potential fields of active nerve and muscle fibers. *Acta Physiol Scand*, 321:15–146, 1969.

[18] Kim G, Ferdjallah M. Non-linear least square optimization of intracellular action potential model using a series of modified gamma distribution functions. In *2nd International Symposium on Computational Modeling of Objects Represented in Images: Fundamentals, Methods, and Applications*, 2010.

7 Robotics

RICHARD LEE

INTRODUCTION

Surgical technological advances have transformed traditional open surgical techniques to minimally invasive surgery.[†] Laparoscopic surgery emerged over 20 years ago. Small two-dimensional laparoscopic cameras and instruments are inserted through small working ports, significantly decreasing patient morbidity. However, two-dimensional imaging limits the surgeon's depth of field, making fine motor movements more difficult to master. Poor ergonomics of fixed-wristed laparoscopic instruments compromised the surgeon's dexterity. These technical constraints restricted advanced laparoscopic techniques to highly skilled and high-volume surgeons and curtailed further progress.

Ten years later these minimally invasive techniques evolved into a new robotics platform. **Surgical robotics** was developed initially at the nonprofit research institute SRI International in the late 1980s. Primary funding was provided by the National Institutes of Health (NIH). A prototype robotics system was developed for its potential to allow surgeons to operate remotely on soldiers wounded in the battlefield. Technical limitations and a time delay in transmission across telecommunication lines had made the performance of surgical procedures from remote locations difficult. Future robotics systems were then developed for commercial utilization.

[†]Minimally invasive surgery: laparoscopic/robotic surgery performed through smaller incisions.

Modeling and Simulation in the Medical and Health Sciences, First Edition. Edited by John A. Sokolowski and Catherine M. Banks.

ROBOTICS TECHNOLOGY

Intuitive Surgical, Inc. is the leader in developing current surgical robotics technology. Intuitive Surgical's **da Vinci system** was first marketed in Europe in 1999.[†] The first system approved by the U.S. Food and Drug Administration (FDA) was the voice-activated camera system Aesop, developed by Computer Motion, Inc. in 1994.[‡] This voice-controlled camera basically freed the surgeon's hand from operating the camera. Computer Motion also developed a robotics platform called Zeus, which was FDA approved in 2001. The da Vinci system was approved by the FDA for general surgery in 2000 and for prostate surgery in 2001. In 2003 these competing companies agreed to merge and the Zeus, was phased out in favor of the da Vinci system.

The da Vinci robotics system is comprised of three components: a surgeon's console, a patient-side robotics cart with four arms manipulated by the surgeon, and a high-definition three-dimensional vision system.

Superior 3D High Definition Vision, an ergonomic surgeon console with Endowrist instrumentation and intuitive motion technology, allows the operator to perform complex procedures with the utmost precision without the constraints of less mobile laparoscopic instrumentation.

Three-dimensional visualization with up to 10× magnification provides a significant improvement from the standard over two-dimensional laparoscopic camera. **EndoWrist technology** provides 12 degrees of freedom with ranges of motion even more than the human hand.[§] The surgeon's precise hand movements are translated and filtered (motion scaling and tremor reduction) to the EndoWrist instruments.

The da Vinci surgical system introduced most recently is the da Vinci Si HD. Several improved features include enhanced high-definition vision, a dual console capability which allows two surgeons working simultaneously to operate the EndoWrist instruments and **TilePro** (a multi-input display which allows the surgeon and operating room team to view the three-dimensional

---

[†]Intuitive Surgical, Inc.: surgical systems developer: together with its subsidiaries, engages in the design, manufacture, and marketing of the da Vinci surgical system. Located in Sunnyvale, California.

[‡]The FDA is responsible for evaluating the safety and efficacy of pharmaceuticals and medical devices in the United States.

[§]EndoWrist technology: Intuitive Surgical's exclusive EndoWrist instruments are designed to provide surgeons with natural dexterity and a full range of motion for precise operation through tiny incisions. Modeled after the human wrist, EndoWrist instruments can offer an even greater range of motion than that of the human hand, with 7 degrees of freedom and 90 degrees of articulation.

display of the operative field along with two additional video sources, such as ultrasound and EKG).

USING ROBOTICS: ADVANTAGES AND DISADVANTAGES

In addition to the advantages that robotic-assisted surgery affords surgeons, there are significant patient benefits as well. Patients may experience less blood loss, improved cosmesis (small incisions with minimal scarring.), less pain, shorter hospital stays, and a faster recovery and return to normal daily activities.

In an era of technological advances, no device has changed the surgical management of certain disease processes as much as the daVinci robotics system. The acceptance of this new technology is unrivaled in such a short period of time. Not only has this major advancement altered the paradigm of surgical teaching, it has made the development of residency training criteria difficult. It is difficult to apply the **Halstedian** surgical teaching model of *see one, do one, teach one* in robotic surgery. There are no standardized guidelines for robotics credentialing. Parameters for safety and competency should be mandatory in the credentialing process. Currently, guidelines for training and proctoring are determined by the individual institution.

Robotics has become a major marketing tool as well. Marketing, combined with the patient's desire for the newest and most innovative procedures, fuels the demand for robotic surgery. The public is often unaware of the potential pitfalls and deficiencies that newer technologies possess in treating their disease. It is obligatory for surgeons and industry to ensure that the entire robotics team is educated in its safe operation and the proper use of this modality.

Potential drawbacks include the significant time commitment for training, the lack of tactile feedback, the financial commitment that hospitals must invest in a robotics system and the significant ongoing maintenance costs. Inherent in robotic surgery is a very steep learning curve and a disconnect for the trainee since the operator is no longer at the patient's side. Acquiring robotic skills during training is a difficult challenge. In an attempt to remedy this dual surgical consoles in the da Vinci Si model allows two surgeons to operate the EndoWrist instruments alternately. The absence of tactile feedback is well compensated for by the significantly improved magnification and three-dimensional stereo optics, which provides visual cues to the surgeon.

The availability of a virtual reality–surgical simulation robotics platform is of extreme importance in improving surgeon proficiency, thereby improving performance and decreasing the trainee learning curve. Integrating simulation into a robotics surgery criterion is not only practical, but may be essential.

Predictive validity is critical since simulation often does not translate into a dynamic surgical environment.

CONCLUSION

Will predefined surgical steps be programmed into newer generation systems, thus bringing us closer to a truly "robotic" system? Possibly the only control the surgeon will have is to make minor adjustments for the dynamic nature of surgery. The robotic platform could also be enhanced by the utilization of real-time intraoperative imaging. This could give the preprogrammed "robot" the ability to make its own refinements. Further improvements would be attained once enough surgical scenarios are programmed into the system. In addition, will the development of reliable high-speed means of data communication improve the latency enough to alleviate the roadblock for telesurgery? A **haptic interface** would also augment the superior visual cues of a robotics system.[†]

Robotic-assisted surgery continues to be an important adjunct to the surgical armamentarium. Appropriate implementation and utilization of this technology is essential to further innovations and its continued safety and success.

KEY TERMS

surgical robotics
Intuitive Surgical, Inc.
da Vinci system
EndoWrist technology
TilePro
Halstedian
haptic interface

[†]Haptic feedbeck: use of the sense of touch in a user interface to provide information to the end user.

8 Training

PAUL E. PHRAMPUS

INTRODUCTION

The process by which a person becomes a practicing health care professional is a long journey that involves various combinations of education, training, and assessment of competence. **Education** can be defined as knowledge acquired by learning and instruction. **Training** cab be defined as activity leading to skilled behavior. Finally, in the context of the health care professions, **competence** can be defined as a proven level of ability demonstrating the application of knowledge and skills in interpersonal relations, decision making, and physical performance consistent with the professional's practice often linked to qualifications issued by relevant professional bodies and in compliance with their codes of practice and standards. It is the combination of education and training combined with a mix of experience that allows one to achieve a state of competency.

Simulation is becoming a part of nearly every aspect of the health care education and training system, spanning from entry-level students through practicing professionals. A variety of factors, are contributing to this, including increases in available technology at reasonable prices, patient safety initiatives that have identified areas for improvement in the current training and education paradigms, and finally, an awareness by system leaders, government officials, and third-party payers of the value that properly implemented simulation programs bring to the health care system.

A major challenge in health care education is standardizing the clinical learning experience that is provided to students. It is becoming increasingly difficult, for a variety of factors, to provide high-quality clinical experience for students involved in health care education programs. Simulation is playing an increasingly important role in the attempt to standardize the clinical experience during training, along with allowing the development of

Modeling and Simulation in the Medical and Health Sciences, First Edition. Edited by John A. Sokolowski and Catherine M. Banks.
© 2011 John Wiley & Sons, Inc. Published 2011 by John Wiley & Sons, Inc.

confidence and the demonstration of competency. This is especially true in the high-risk, low-frequency events that students typically would not be able to manage independently, as well as low-risk routine skills that have not had robust standardized rubrics of competence applied previously.

Many factors are responsible for a proliferation of simulation training programs. Patient safety is among those that are most important. An acknowledgment of the number of patient deaths attributable to health care system error was published in 2000 and raised public and lawmaker interest in creating a safer health care system in the United States [1]. The traditional teaching apprenticeship model, which involves performing medical procedures on actual patients, is being preempted by requiring a demonstration of competence in the simulated environment. For example, simulation-based mastery training programs have recently been demonstrated to decrease infections, complications, and the cost of caring for patients who undergo complicated procedures such as the insertion of a central venous catheter [2–4].

The utilization of simulation in terms of styles, methods, technologies, and processes varies significantly. They range from the use of highly sophisticated computerized replicas of human bodies that respond and provide feedback similar to the responses expected from a human being, to that of a simple plastic arm as model to provide education in starting an intravenous line into the vein of a patient. Similarly, the environment in which simulation training programs occur can range from a lifelike depiction of an operating room or battlefield to that of a standard hotel conference room filled with simulation equipment with no attempts at recreating a medical environment. The learning and assessment objectives should drive the decision making needed for the equipment and environmental fidelity that is required to create a successful program.

The stages of a person becoming a practicing professional can be broken down into three phases: student, new graduate or graduate student, and then practicing professional. The phases are named slightly differently depending on the discipline of health care, but from an education and training perspective this categorization proves useful as a means of discussion. The learning objectives and expectations vary significantly depending on the phase of the learner participating in the simulation event.

Students are more likely to be engaged in singly focused learning objectives as building blocks for a more complicated integration and analysis of the data they are learning to obtain in the early phases of their education. Later in the student phase they will be expected to aggregate the knowledge that they have obtained in the simple building blocks and to analyze and make decisions regarding treatment plans that demonstrate an interpretation of the data they have obtained.

This layer of integration and assessment and analysis is the core focus of the new graduate phase, or in the training of physicians, what is known as the residency phase. Practicing professionals need to engage in a lifelong process of learning the most current therapies, procedures, and best practices, which change rapidly with evolving health care literature, and be able to demonstrate continued competence.

Heretofore, the competence of many involved in the provision of health care has relied extensively on knowledge-based testing by the use of multiple-choice questions and other standardized examinations that focus on the acquisition of one's accumulated body of knowledge. Unlike knowledge-based testing, the actual determining of competence from a performance perspective has not reliably been nearly as robust or structured. Performance in the simulation environment can enhance some of the standardization of assessments of competence in cases where the technology exists to provide the necessary **face validity.**

Most models of education programs of health care professionals involve heavy doses of didactic learning, reading, and workshops combined with clinical experience. The current model of clinical experience is known as "structured education by random clinical opportunity," meaning that we rely on the random chance that the student or new graduate will experience a situation to complement the didactic learning while in the clinical environment. Simulation has the capability to standardize aspects of one's clinical experience to ensure that the exposure and immersion is one of a more structured opportunity. This is particularly important now, as changes in the practice of health care are making it increasingly difficult for training and education programs to have good-quality, prolonged, and immersive clinical experiences. Factors such as increases in the numbers of students enrolled in health care programs in response to workforce demand, medical legal concerns, and clinical productivity requirements of teaching faculty are some examples of challenges to the provision of good-quality clinical experiences.

The transition phase between new graduate and practicing professional varies widely among different domains of health care provider. Typically, in the area of nurses and paramedics, once students graduate from an accredited program, they enter a period of supervision in which they are paired with a "clinical preceptor" and undergo varying degrees of observation of their clinical performance prior to being released in the clinical environment as an independent practitioner. In some cases there are continued knowledge-based evaluations and written tests during this precepting time frame, but often the clinical performance evaluations remain unstructured and lack standardization. Simulation can help to standardize this transition phase as well as help to reduce the period of requiring the learning of new and additional skills to be functional at an independent level. This period, known as the **practice gap**, is the difference between what new graduate professionals need to complete accredited training programs and the knowledge and skills that they need

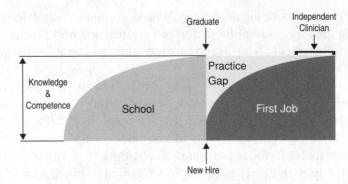

FIGURE 8.1 Practice gap.

to acquire to be independent practicing clinicians (Figure 8.1). Recent esti-
mates calculate that hospitals spend over $100,000 preparing a newly hired,
new graduate nurse to become a fully functioning independent health care
provider [5–7].

The training pathway leading to that of a board-certified physician involves
a residency program after graduating from medical school that ranges from
three to seven years in duration. During this phase the resident physician
undergoes multiple knowledge-based testing along with observations of performance in the clinical practice environment. Whereas the knowledge-based
evaluations are structured and standardized, the actual clinical performance
evaluations vary widely and are for the most part without standardization.

The rapid changes that now occur in the body of health care literature and
the increasing amount of technology that is being brought to the forefront
of patient care are demanding new methods of training and assurance of
competence such as at no other time in history. New emphasis on quality
in health care, reducing errors, and patient safety and cost efficiencies of
the provision of health care brought forth by government regulators, private
payers, and consumers of health care are demanding new ways of ensuring the
proficiency of individuals as well as teams of people who care for patients in
the twenty-first century. Simulation has been recognized as being an important
part of this future [1].

Factors influencing the incorporation of simulation into health care education are similarly just as varied. Standardization, funding, availability of
technology, faculty development, and acceptance by students and faculty
members are just some of the considerations that factor into the implementation of simulation systems in health care training. Despite these challenges,
the incorporation of simulation in health care training programs is increasing
exponentially around the world. In 2005 there were fewer than 50 known simulation centers throughout the world. By 2010 there were hundreds, perhaps

thousands, of installations of simulation-based training programs across all domains of health care as well as all levels of trainees, from the student level to that of the practicing professional.

TYPES OF TRAINING

While numerous subcategories of different types of training are needed to address the entire spectrum of health care providers, they can be broadly categorized into five groups (Table 8.1): (1) communications and interviewing skills; (2) psychomotor skills where the necessary mechanics of accomplishing a specific technical procedure are covered; (3) decision making; (4) combinations of psychomotor skills and decision making; and (5) team training, in which the contributions of a team, or a person's contribution to a team, can be taught and assessed. With regard to team training, the difference is that the focus is less on the medical content and much more on how teamwork skills, presentation skills, and the interactions between team members can contribute to the care of a patient.

Communications and interviewing skills involve training and assessment that revolves around a trainee's ability to communicate effectively with a patient, or family members of a patient, in an attempt to glean medical data on which to help base decisions. In addition to assurance of the ability of health care students to be compassionate in their discussions and assist the patient family with decision making, there is an important underpinning of proper medical interviewing that allows for creation of an accurate diagnosis and eventual treatment plan. Medical interviewing and history taking is a very intricate process because of the branching in decision making that occurs in response to the information garnered from the interview and physical examination process.

Psychomotor skills training is concerned with the mechanical performance of technical procedures or physical examination techniques. This can range from simple tasks such as establishing a peripheral intravenous line in a patient to an extraordinarily complicated laparoscopic surgical procedure,

TABLE 8.1   Broad Categories of Training

Broad Categories of Training
1. Communications and Interviewing
2. Psychomotor Skills
3. Decision Making
4. Combination of Skills and Decisions
5. Team Training

such as removing someone's gallbladder through a series of small incisions. Psychomotor skills training incorporates some review of the student's knowledge of the procedure, indications, contraindications, equipment, patient's anatomical landmarks, potential complications, and steps necessary to perform the task. The focus of the actual skills training, however, is to develop in a methodical way the motor skills necessary to carry out the procedure in a safe and effective fashion.

Simpler procedures rely on task trainers: for example, silicone or latex intravenous arms in the case of intravenous lines. Variants of simulated real tissue are used as well. For example, in training of the placement of an intraosseous, which essentially embeds a large-bore stainless steel catheter into the leg bone of pediatric patients, many training programs use chicken legs and turkey legs as the simulated tissue to increase the **fidelity** of the simulation because proper placement can result in the aspiration of blood and bone marrow tissue, closely resembling a real-world clinical example.

For more complicated task training such as that associated with performing colonoscopy or laparoscopic surgical simulations, a number of different training devices have emerged. Some are very simple systems that have a student demonstrate basic competence with psychomotor skills such as cutting with scissors while watching a screen. Basic skills have been validated that are thought to transfer to real-world competence in dealing with the equipment associated with procedures.

Other more complex systems are available, such as computerized systems (Figure 8.2) that combine aspects of actual surgical instruments, virtual screen presentations based on computer modeling as well as systems that incorporate haptic feedback, varying levels of complexity, and scoring capabilities. In many of these systems, in addition to motor skills training, there is an incorporation of the decision making built into the system. For highly complex and detailed surgical procedures, live tissue, or live animal labs, are often utilized for competency development, as there is still a technological void in this area.

Decision making is an important part of providing care to a patient. Data need to be gathered from a variety of sources, integrated, and analyzed, and from this analysis a decision needs to be made to craft a plan of care for the patient. Although there is some level of decision making associated with every procedural skill described above, simulations involving decision making are more complicated because there are usually many correct but different pathways leading to the proper care of the patient. Thus, simulations that focus on decision making must provide the background environmental milieu for trainees to base their decisions.

Combining the skills and decision-making types of training is much more common and closer to realistic care of the patient. It is often difficult to

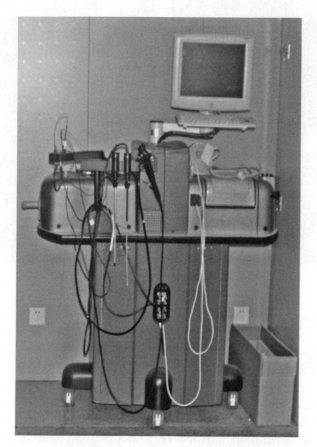

FIGURE 8.2   Computerized simulator.

combine high levels of psychomotor skills evaluations with an analysis of decision-making skills in a given situation. Additionally, it is a far more effective stepwise process to allow the training to master the skills portion and then add the decision making as another layer of complexity in the training pathway.

As the level of the sophistication of the health care provider increases, there are more overlaps in the type of simulations that are employed. For example, in the initial training of students, it is very common to focus purely on a specific task-training procedural mastery. Once baseline competence is demonstrated in the procedural task, a communications element may be added. For example, a student may undergo rigorous task training to learn how to suture a wound using a partial task trainer such as a pig's foot or other surrogate for human skin. In later training, a hybrid simulation may be employed in which a task trainer that represents a wound in a patient's skin may be strapped to the simulated patient. In this situation, a combination,

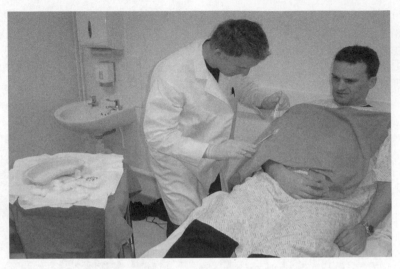

FIGURE 8.3   Hybrid simulation.

or **hybrid simulation**, can occur that allows an assessment of the ability to perform the task combined with an assessment of the skills of the trainee to communicate with the patient in an acceptable manner (Figure 8.3).

Team training focuses on the roles and goals associated with people who possess complementary skill sets and come together to provide patient care. Interpersonal communication skills, effective communication strategies, situational awareness, and team leadership are often the focus of such training encounters. In general, there is less emphasis on individual decision making and more on teaching the participants that teamwork functionality alone can contribute to the effectiveness of patient care. There are many different strategies associated with team training, many of which were rooted in outside industries such as the aviation industry and nuclear power.

An example of a team training curriculum, developed by the Department of Defense and Agency for Healthcare Research and Quality, is the Teams Stepps Program, which is an evidence-based teamwork system to improve communication and teamwork skills among health care professionals.

CREATING SUCCESSFUL SIMULATIONS IN HEALTH CARE

As mentioned previously, there is a wide variety of successful implementation of simulation methods in training programs. It is recognized that there are many possibilities of how to conduct simulation-based training programs, and at present, with a lack of standardization, it is foreseeable that many more methods will emerge.

Successful simulation training programs begin with a careful evaluation of learning objectives and a consideration of the assessment necessary to achieve the learning objectives and demonstrate competence. It is important that the simulation training program not develop around a specific technology just because the technology exists or provides a given feature. A careful analysis of the learning objectives, and then an evaluation of the available resources, will be combined to create effective submission-based training programs.

Some guiding principles or "best practices" have been published by several authors over the last several years to assist in guiding the creation of the most effective programs. One of the first to attempt to quantify successful attributes of embedding simulate high-fidelity simulation into health care education was published in 2005 as part of a Best Evidence in Medical Education produced by the Association of Medical Education in Europe [8]. The association identified program design features associated with effective learning utilizing high-fidelity simulation:

- Feedback is provided during learning experience.
- Learners engage in repetitive practice.
- Simulation is integrated into the overall curriculum.
- Learners practice with increasing levels of difficulty.
- The method is adaptable to multiple learning strategies.
- Clinical variation is provided.
- The environment is controlled.
- Learning is individualized.
- Outcomes and benchmarks are clearly defined.
- The simulator is valided.

To create the most successful simulation-based training program or event, careful consideration needs to be given to the level of the learner; the logistics, such as the amount of time students have dedicated to the exercise; and the resources available. Resources include not only the availability of technical simulation equipment but also that of the technical personnel, administrative support, facilities, and properly prepared simulation-based faculty members needed to conduct the training.

Once the level of the learner, the resources, and the learning objectives have been analyze and defined, the program can be designed. It is the reflection of the learning objectives of the educational encounter that should drive decision making during the creation of a program. The analysis of available resources, combined with the expectations of the student, should be considered carefully. Since one of the rate-limiting steps of the health care education process is

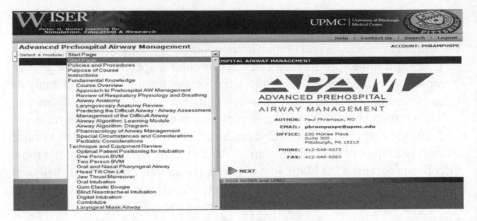

FIGURE 8.4   Precourse material.

faulty members, it is important to regard the time that the students are in proximity of a faculty member as highly valuable. In addition, facilities that provide simulation-based education should maximize the amount of hands-on activities that occurs and place less emphasis on lecturing.

Many simulation programs have adopted a strategy of having the learner prepare for the simulation event(s) by refining or maximizing their cognitive knowledge regarding the topic prior to arriving at the simulation event. This can be satisfied by assigning prereading or other asynchronous learning modalities, such as Web-based education modules, which allow for interactive learning (Figure 8.4). It is a more effective use of the simulation faculty, equipment, and facilities if the student can be prepared prior to the simulation activities through combinations of participation in multimedia education strategies as well as pretesting and self-assessment. This allows a preservation of the time spent in the highly valuable environment of the simulation environment with faculty members to that of maximum effectiveness. This strategy is employed at the University of Pittsburgh's simulation center via a Web-based learning system that is associated specifically with their simulation-based training programs.

Successful simulation training programs require extensive preparation prior to the event. Attention to detail and logistical considerations in a given simulation exercise are germane to the success of the program. The training program must be organized thoughtfully and be reflective of the goals of the learning objectives, combined with a consideration of the logistics of actually accomplishing the program.

Properly oriented learners who are aware of the goals and objectives of the program are usually willing to engage in a state of suspended disbelief that allows the student to treat the simulations realistically. The simulation program needs to have a thoughtful schedule and carefully constructed scenarios that afford an adult learner the realization that they are engaging in relevant learning activities that are appropriate for their level of education and are likely to be important in their future or current role as care providers of patients.

Simulation scenarios are the building blocks of simulation-based education programs. Scenarios contain background information to acclimate the learner with the environment that is being portrayed, a level of environmental fidelity and/or physiological fidelity portrayed by the simulator that will allow the learner to engage in the process of data gathering, analyzing, and creating decisions that will be evaluated by the assessment objectives associated with the scenario.

Once the learning objectives have been achieved or other predetermined set points in the scenario have been reached, the scenario will end. Usually, a **debriefing** will follow, during which faculty members and trainees will engage in a thoughtful review of the scenario with a discussion emphasis based on the specific objectives for which the scenario was executed.

As only a finite amount of time is available for education encounters, it is important that the debriefing leader, usually a faculty member, constrain the discussion of performance to those who are targeted in the objectives to run the scenario. Depending on the overall goals of the scenario (e.g., focusing on psychomotor skills, on decision making, or on teamwork), there are various methods of debriefing available to accomplish the objectives. Often, a chronological review is undertaken of the performance that occurred during the scenario, and thought-provoking discussions are prompted by the person leading the debriefing, in accordance with the overall objectives in conducting the scenario. In some cases, when pointing out adherence to or variation from standardized treatment guidelines, national best practices or treatment algorithms may be discussed as well as the rationale or decision making that led to the variance.

Most practioners believe that the debriefing is the crucial time for the reinforcement of knowledge and correction of errors to occur through a guided process that allows reflective learning. Thus, in faculty development, attention to development of skills and methods of debriefing is crucial.

Some simulators provide embedded technology to assist in the general scripting of the debriefing by keeping a running-time-stamped log of the major events that occurred, a video recording of the event, and the ability to incorporate performance-specific, contextually based commentary that promotes a rich discussion between faculty and trainees. This type of embedded technology also assists in keeping the script of the debriefing to a fairly narrow window that remains close to the intended learning objectives.

Scenarios also need to contain logistical information regarding faculty members and their role in the scenario. Such logistical information includes props, supplies and equipment, and the simulation room setup necessary to conduct the training. Finally, a number of administrative considerations contribute to the success of simulation-based training programs, such as simulation room scheduling, faculty recruitment, and timely reminders.

Coordinating the overall design of a simulation program requires a careful balance of theory, best practices, and the constraints of reality. The constraints are represented by various factors, such as the availability of personnel and equipment, the amount of time that students have in which to participate, and in some cases the facilities in which the training is conducted. Practicing professionals often represent the category of learners with the most limitations on time.

Given these limitations combined with various learning objectives, many styles of simulation deployment have emerged. Scheduling students for a block of time to be used for a simulation-based course is one such method. In this case groups of students typically participate in a number of related scenarios arranged around a central learning theme.

Other methods include scheduling a single scenario to cover specific objectives. In this case, no substantial commitment of time may be required for the entire course. **In situ simulations** are becoming more common and are conducted in proximity to, or in, the actual environment in which real patient care occurs. Advantages include increasing the level of environmental fidelity as well as the fact that practicing professionals do not have to be away from the patient care location for extended periods of time. Disadvantages include real patient care obligations that can interfere with the training program, and in the case of actual patient units, the availability of space to conduct the simulation.

TRAINING PROGRAMS AND STUDENT LEVELS

Training programs in simulation have varying needs in terms of the curriculum that is constructed, the fidelity levels that are required, and the tasks that are expected to be completed based on the level of the learner. A wide spectrum of equipment is employed in the use of simulation-based training programs. A discrete list of the type of agreement used at the student level is not possible because of the creativity involved in creating such training programs. However, there are some generalities associated with training programs are created for each level of students.

Students can broadly be characterized into four major learner categories: students of a particular domain of health care, graduate students, entry-level professionals, and practicing professionals. The learning objectives and the needs of each vary considerably, and this has important implications on the

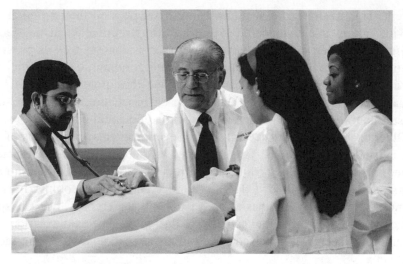

FIGURE 8.5   Harvey cardiopulmonary simulator.

curriculum design as well as the equipment that is employed to carry out the simulation encounters.

Medical and nursing students are often involved in learning highly segmented detailed tasks or discrete learning objectives that will later be amalgamated into higher-level processes such as the complete history and physical examination of the patient, versus the task of taking a blood pressure. Student-directed simulation-based learning programs often use many partial task trainers to accomplish mastery in specialized aspects of clinical practice. For example, the Harvey cardiopulmonary simulator has been well validated to be able to assess competence in a student's ability to perform a cardiopulmonary examination to allow for the recognition of clinical signs and to be able to assess a student's ability to recognize heart murmurs (Figure 8.5).

Graduate-level students are employing higher-level assessment and decision making into their practice. They have previously demonstrated mastery of the student level. Graduate-level students are often engaged in treating more complex cases, performing more complicated procedures, and having greater resource knowledge, as well as developing competency in their role as a member of the health care team.

Many specialty graduate-level programs exist to assist with the efficiencies of education. Examples of graduate-level programs include demanding that a minimal level of competencies be achieved prior to stepping forth and engaging in actual clinical practice. Residency programs in anesthesiology have successfully employed competency-based simulations which ensure that junior residents have a minimum basic competency in the skill set that allows them to entering the operating room and administer anesthesia to a surgical patient. This assurance of a minimum skill set prior to their being able to provide care in an actual operating room allows the clinical teaching faculty to focus on higher levels of education for the resident, as well as providing important aspects of patient safety and operating room utilization efficiency. Previously, even the minimum basic skills of anesthesiology were taught in the actual operating room during an actual surgical procedure, resulting in longer anesthesia times for patients and the associated potential complications.

Other examples of advanced graduate-level programs are in areas of sub-specialty clinical rotations. For example, at the University of Pittsburgh, upper-level anesthesiology residents are required to pass a simulation-based course demonstrating knowledge and competence in the complexities and nuances of providing anesthesia care of patients undergoing liver transplant procedures prior to providing actual care during this specialty rotation. This program has been well received by residents as well as by teaching faculty members.

Entry-level professionals are those transitioning from training programs into actual practice. This differs by domain. Whereas nurses typically grad-uate from nursing school directly to clinical practice, the education path-way of physicians requires that the graduate attend a residency or graduate-level training program before transition into actual unsupervised clinical practice.

Entry-level professionals are now beginning to undergo standardized simulation-based education programs to ease the transition to that of a practic-ing professional. Types of simulation include orientation to the environment, the role of the professional in a particular specialty environment, competency in procedures and tasks associated with specialty areas of the clinical envi-ronment, and orientation to the clinical practice team associated with one's workplace.

Examples of such entry-level simulation-based programs include operating room nurse orientation programs and emergency nurse training programs that orient nurses who are new to the operating room and emergency department environments. This specialized type of simulation-based training is believed to standardize the orientation approach, resulting in cost savings for personnel as well as achieving higher levels of patient care.

Simulation-based training program for practicing professionals are becoming more common. Training programs for practicing professionals are generally designed to refresh skills and to ensure currency and competency of knowledge and modernization of their overall practice. Recently, the American Board of Anesthesiology has set a requirement that all recertifying anesthesiologists will have a component of simulation-based training as part of every certification process. Within the practice of anesthesiology, this program is known as the MOCA (Maintenance of Certification in

Anesthesiology) program. Simulation centers that have successfully gone through the American Society of Anesthesiology accreditation process are authorized to conduct MOCA training programs that count toward a physician's recertification continuing education requirements.

SIMULATION-BASED TRAINING

Incorporating Assessment

Successful simulation programs can span the continuum of events created purely for education to that of simulation scenarios that are used specifically for assessment of competence. Education scenarios may be used to enhance the interactivity, reinforce complicated points, or sometimes to create relevance to the student of seemingly esoteric information.

Some programs have embraced these capabilities of the use of simulation and create programs that contribute to both the educational mission and assessment. The development of such programs as well as the faculty support needed to carry them out are critical aspects of faculty training that are specific to simulation-based education.

Although traditional educators and health care are comfortable with the concept of lecturing and running workshops for skills development, very few have been engaged in programs that provide education and nearly simultaneous feedback as well as incorporating assessment capacity. The traditional model has been the dispensing of information, reading assignments by students, and then assessing knowledge at the end of a given learning period or, commonly, a semester. This traditional pathway relies heavily on the assessment of didactic knowledge through a standardized multiple-choice written test, and does not effectively evaluate performance or competence or the ability to put the knowledge to use in the setting of care for patients.

Highly effective programs that involve simulation incorporate aspects of education along with making observations of competency in analysis, performance of skills, decision making, and team aspects. Many programs involve multiple simulation scenarios to allow for repetitive practice for mastery learning as well as a platform to correct deficiencies and other scenarios carriedout for assessment of competence.

In the domain of medical education, standardized board examinations for medical licensure in the United States have incorporated various aspects of simulation-based activities in the form of an OSCE (objective structured clinical examination) section that has been incorporated into licensing exams [9]. In Canada, all physician candidates applying for licensure are required to demonstrate cardiopulmonary assessment skills on the Harvey simulator mentioned previously. This infusion of simulation allows the evaluation of some levels of performance in addition to the traditional paper-based testing of knowledge.

Fidelity in Simulation Training

As related to simulation, the word *fidelity* may have as many definitions as may any other word. In general, fidelity can be defined as how close something represents reality. When simulation is used in the training of health care providers, there can be many definitions. The *fidelity of simulation equipment* may represent how close the equipment performs relative to the anatomy or physiology of a human being. This can be in terms of how lifelike an anatomical structure appears, or perhaps how it feels or responds to external stimulus. Similarly, *physiological fidelity* refers to how realistic the physiology of the simulator is in response to the treatment(s) provided by the student.

Little is known regarding the effect of fidelity on the actual outcomes of the learning achieved with simulation-based health care training. With regard to anatomical and physiological fidelity, educators have relied on the technology that has been commonly available from equipment manufacturers.

Environmental fidelity refers to a comparison of the setting that the simulator is in compared to that of the health care environment in which the actual care takes place. Examples include the operating room, the emergency department, or the back of an ambulance in the case of training for prehospital care providers. Environmental fidelity varies tremendously among established simulation centers, with some centers spending extraordinary amounts of money and time to re-create a nearly lifelike and functional environment that mimics that of the actual health care space. Some programs simplify the amount of fixed infrastructure that re-creates the real environment in favor of a more modular design that allows accommodation for participants of varying clinical backgrounds.

The needs for the expenditure and planning associated with increasing environmental fidelity are determined by the learning objectives. For example, if the learning objectives of a helicopter-based transport team are for the team to demonstrate competence in the management of a difficult airway situation according to established protocols, the simulations can be carried out in an environment as simple as a hotel conference room. In contrast, if the learning objectives required further that the demonstration of the competence and management of the patient occur in the austere environment of the helicopter, which includes cramped spacing and the inability to utilize pulmonary auscultation methods because of noise constraints, the environment of the simulation would need to closely mirror that of transporting a patient in an actual helicopter.

Equipment Used in Simulation Training

The equipment used in simulation varies widely. Ideally, the equipment selected would be for a pure evaluation of the learning objectives and assessment strategy for a given topic. However, the reality of financial resources, trained faculty members, the amount of time that can be dedicated to the simulation activities, and the physical space available often determine the final equipment lists that are incorporated into simulation-based training programs. Examples of basic categories of equipment include task trainers, computerized test trainers, computerized human simulators, virtual reality simulators and standardized patients.

Task trainers are representative replicas of a small piece of anatomy of a human being that is equipped to function, providing relative anatomical structures to allow the performance of a technical procedure. A plastic airway head used for intubation training is a classic example of a task trainer.

Computerized task trainers are another level of complexity. They are often augmented with sensors and perhaps even haptic feedback that allow the learner to navigate a virtual environment while holding the actual instruments in their hands. A typical example of this is a laparoscopic surgical simulator.

Human simulators are a category of equipment that in general provide a replica of an entire human body and often contain computerized and pneumatic equipment along with physiologically based algorithms or the ability to control the physiologic state that is portrayed to the student to allow the recreation of the state of the human being to accomplish the learning objectives of the given scenario. In addition to robust physiologic findings such as vital signs, diaphoresis, and hemorrhage, for example, task training functionality is embedded into the simulator to enable the performance of medical procedures as well.

Human simulators also generally embed the ability to record the physiological state of the simulator and often records of the major events that occurred during the simulation that are to be incorporated into the debriefing session. Human simulators also tend to have various foci on actual human qualities, simulated depending on the implementation intended. Some are very broad-based, such as those used for emergency and resuscitation cases; others are specific, such as those used for the training of childbirth.

With regard to the functionality of human simulators, it is important to realize that an exact replica of a human model would not be the most ideal tool to use in a simulation scenario. The best simulation tool to use in a health care scenario is one that allows the portrayal of the simulated clinical encounter but, most important, that allows the learning objectives to be accomplished. While aspects of physiology-based modeling have been incorporated successfully into some simulators used for training programs, it is important that the design still incorporate levels of flexibility and control that can be manipulated artificially to meet the needs of the training environment.

Many centers being developed today spend inordinate amounts of money and space trying to create an exact replica of the health care environment, thinking that this increases the environmental fidelity. However, it is important to remember that the goals and objectives of a hospital are to care for patients, and thus design decisions are made respective to that mission. When designing simulation centers, it is important to realize that decisions revolve around the fundamental mission of education, not of health care.

Simulations involving communication and interviewing skills often employ the use of **simulated** or **standardized patients**, actual highly trained human beings who portray the role of a patient with a specific disease process. Often, these encounters include observing students mastering physical examinations and interviewing skills on simulated patients. Although there are emerging technologies with virtual interviewing, and virtual patients who allow a two-way dialogue between a health care provider and a virtualized patient, so far they have not been able to overcome the cost prohibition or to achieve acceptable fidelity to be placed into widespread use. The use of simulated patients and health care education and assessment is in widespread use, and in many cases is used for high-stakes assessments such as passing licensure exams. Standardized patients have been validated to be able to be trained to report accurate assessments of the simulated encounter [10].

Other equipment used in educational simulation includes standard PCs, and sometimes specialized equipment for Web-based simulations, tabletop exercises, and teleconferencing equipment that allows the learner and faculty member using the simulation to be separated in space.

The next category of equipment resource to be considered is the operational infrastructure used to support the simulation-based education environment. In contrast to traditional educational settings, where a projector and screen were all that were needed, many simulation centers today are evolving as highly technologically specialized environments.

Props are a category of equipment that is essential to good-quality simulations. These are devices, objects, or actual equipment in proximity to the physical simulation that create a sense of environmental reality to the training participants. In addition to props, other specialty equipment may be needed, depending on the level of environmental fidelity that is to be achieved. For example, specialty lighting equipment such as strobe lights, audiovisual equipment that can generate sound effects, and machines to produce smoke and fog may be incorporated into the design of a training program. Careful cross-referencing to the goals and objectives of the program is always needed when contemplating the decision as to what to include and what not to include. Each inclusion in the scenario increases the complexity of conducting the training program and, in some cases, creates more opportunity for technological glitches that can interrupt the flow of the scenario.

Audiovisual capturing systems are commonly incorporated into simulation-based training programs. Many programs involve audio and video recording of the simulation scenarios to be used as part of the feedback strategy to be covered in the debriefing. Specialty camera and audio recording equipment that allows multiple views of the environment where the simulation is occurring is becoming commonplace in the design of new centers. It is particularly helpful to be able to identify specific times in the simulation where a discussion point or learning objective was achieved, or where best practice was followed or varied from, and to be able to recall that time frame. These are examples of important points of the simulation that may be incorporated into the debriefing to help with the guided reflection of the student(s) who are undergoing the debriefing. Other strategies of debriefing are predicated on watching segments of the video and reviewing the performance of each team member individually with respect to assigned roles and goals in a particular scenario.

Important considerations in the design of simulation-based training programs are decisions surrounding the audiovisual capture of the performance of the scenarios, including trainee's preference not to be recorded, and legal and liability issues associated with distribution of the recordings. This can be a significant consideration in the design of programs, as the most common reasons that practicing nurses are reluctant to participate in simulations is related to being videorecorded [11].

Recording and archiving of simulation performances also has other operational, technical, and administrative considerations. During the course of simulation scenarios, large volumes of data can be amassed. This is true in the case of stored videos documenting the scenarios. Careful administrative oversight of the recording data is an essential part of planning simulation training.

When considering a simulation-based training program it is important that the administrative as well as operational aspects of the information technology services available to support the data be considered.

BARRIERS TO SIMULATION TRAINING IN HEALTH CARE

Widespread adoption of simulation-based training across the health care education community has been embraced with varying degrees of acceptance, implementation, and incorporation into the fundamental education training programs. There are common barriers to transcending the domains of health care with regard to the implementation of simulation.

A lack of standardization of the role of simulation throughout health care education is a principal barrier to its adoption. Thus far there has been little research or a commitment to sets of standards involved with the implementation of simulation-based training. Many different providers of health care education are creating simulations and utilizing them in different methods. This probably contributes to the systematic slowing of adoption.

One of the most significant rate-limiting steps in the widespread adoption of simulation is a lack of adequately trained faculty members who are confident in the use of simulation-based education techniques. A faculty member is often the rate-limiting step to successful simulation efforts. Simulation-based training is essentially small-group learning. This necessitates lower faculty-to-student ratios for simulation-based activities than for safer traditional-based educational activities such as lecturing.

Simulation-based training is often thought of as expensive and inefficient because of the small-group learning aspect, combined with the need for increased numbers of faculty and a perception of expense associated with equipment, technical resources, and adequate facilities to conduct the simulation training. Another barrier to simulation-based training is a reluctance of education leadership as well as many faculty members to embrace new methodology and training systems associated with its use.

KEY TERMS

education	psychomotor skills	task trainer
training	hybrid simulation	human simulator
competence	fidelity	simulated or
face validity	debriefing	standardized patient
practice gap	insitu simulation	

REFERENCES

[1] Kohn LT, Corrigan J, Donaldson MS. *To Err Is Human: Building a Safer Health System*. Washington, DC: National Academy Press, 2000, p. 287.

[2] Barsuk JH, et al. Use of simulation-based education to reduce catheter-related bloodstream infections. *Arch Intern Med*, 169(15):1420–1423, 2009.

[3] Barsuk JH, et al., Simulation-based mastery learning reduces complications during central venous catheter insertion in a medical intensive care unit. *Crit Care Med*, 37(10):2697–2701, 2009.

[4] Cohen ER, et al., Cost savings from reduced catheter-related bloodstream infection after simulation-based education for residents in a medical intensive care unit. *Simul Healthcare*, 5(2):98–102, 2010.

[5] Burns P, Poster EC. Competency development in new registered nurse graduates: closing the gap between education and practice. *J Contin Educ Nurs*, 39(2):67–73, 2008.

[6] Dyess SM, Sherman RO. The first year of practice: new graduate nurses' transition and learning needs. *J Contin Educ Nurs*, 40(9):403–410, 2009.

[7] Marshburn DM, Engelke MK, Swanson MS. Relationships of new nurses' perceptions and measured performance-based clinical competence. *J Contin Educ Nurs*, 40(9):426–432, 2009.

[8] Issenberg SB, et al. Features and uses of high-fidelity medical simulations that lead to effective learning: a BEME systematic review. *Med Teach*, 27(1):10–28, 2005.

[9] Dillon GF, et al. Simulations in the United States medical licensing examination (USMLE). *Qual Saf Health Care*, 13(Suppl. 1):i41–i15, 2004.

[10] Harden RM, Gleeson FA. Assessment of clinical competence using an objective structured clinical examination (OSCE). *Med Educ*, 13(1):41–54, 1979.

[11] Decarlo D, et al. Factors influencing nurses' attitudes toward simulation-based education. *Simul Healthcare*, 3(2):90–96, 2008.

FURTHER READING

Aldrich C. *Simulations and the Future of Learning: An Innovative (and Perhaps Revolutionary) Approach to e-Learning*. San Francisco: Pfeiffer, 2004, p. 282.

Aldrich C. *Learning by Doing: A Comprehensive Guide to Simulations, Computer Games, and Pedagogy in e-Learning and Other Educational Experiences*. San Francisco: Pfeiffer, 2005, p. 353.

Fink LD. *Creating Significant Learning Experiences: An Integrated Approach to Designing College Courses*. San Francisco: Jossey-Bass, 2003, p. 295.

Loyd GE, Lake, CL, Greenberg RB. *Practical Health Care Simulations*. Philadelphia: Elsevier Mosby, 2004, p. 613.

Riley RH. *Manual of Simulation in Healthcare*. New York: Oxford University Press, 2008, p. 548.

Tekian A, McGuire CH, McGaghie WC. *Innovative Simulations for Assessing Professional Competence: From Paper-and-Pencil to Virtual Reality*. Chicago: University of Illinois at Chicago, Department of Medical Education, 1999, p. 254.

9 Patient Care

EUGENE SANTOS, JR., JOSEPH M. ROSEN, KEUM JOO KIM,
FEI YU, DEQING LI, ELIZABETH A. JACOB, and
LINDSAY B. KATONA

INTRODUCTION

Until 1999, when the seminal report *To Err Is Human* was announced, most health care professionals were unaware of the significance of their own mistakes, because there was little or no supporting infrastructure to report and track adverse events and medical errors [1–3]. Since then, a wide range of research has been conducted to identify risks to patients and to prevent medical errors. Although the research has significantly enhanced patient safety, adverse events still occur occasionally as surgical procedures become more complex through the development of advanced medical technologies and our extended life expectancy. According to a report by Pronovost et al., the most common errors were related to medication (42%), incorrect and incomplete delivery of care (20%), and equipment failure (15%) [4]. For example, medications may be ordered incorrectly, due to a misinterpretation of a doctor's handwriting on a prescription; medical errors may be attributed to a failure in administering the correct dosage during the appropriate time frame; incorrect and incomplete surgeries may be performed due to retained foreign bodies, wrong-site operations, mismatched organ transplants, and incompatible blood transfusions; and equipment failure may occur when the required instruments are arranged incorrectly, sterilized improperly, or mismatched. These errors seem inevitable when we consider medical practices in which complicated and critical decisions are necessitated, frequently with limited and conflicting information [5,6]. In addition to this, ensuring patient safety becomes even more challenging when a patient is transferred from one institution to another or from one surgeon to another. These medical hand-offs happen frequently both in the theater of war and in civilian medical practices [7–10].

Modeling and Simulation in the Medical and Health Sciences, First Edition. Edited by John A. Sokolowski and Catherine M. Banks.
© 2011 John Wiley & Sons, Inc. Published 2011 by John Wiley & Sons, Inc.

The OR (operating room) is particularly susceptible to medical error, due to its complex and multidisciplinary nature. Communication among team members becomes more difficult, as shown in a report stating that over 70% of sentinel events are associated with teamwork and communication in obstetric critical care [11].[†] This is also indicated by a study reporting that even the level of teamwork in the OR is perceived differently by team members such as surgeons and nurses [12]. Medical errors rooted in miscommunication generally occur when team members happen to have different viewpoints [13]. Doctors may change orders without adequately communicating with nursing staffs; incorrect patient information may be passed through different medical teams; team members' responsibilities may be delegated ineffectively and their roles may not be clarified in detail; or some team members may have inaccurate assumptions of the knowledge and skills of other members. Although a wide variety of factors contribute to the high risk of medical errors in the OR, poor and ineffective communication is a major contributing factor [14]. In addition, the Joint Commission on Accreditation of Healthcare Organizations has identified communication breakdown as a leading cause of medical errors [15]. There is a strong consensus that communication in the OR is essential to patient safety and quality care [12,16,17]. Up until now, however, findings are still very limited and more research is necessary. To that end, we provide a computational methodology to improve communication among medical team members in the OR by analyzing gaps while inferring intent of the team members. We assume a system that monitors the OR team members continuously and aims to assist their understanding of dynamic situations/environments and of their co-workers in order to enhance patient safety and the quality of medical care.

Communication and information sharing are essential to patient safety but can easily break down in the OR with people of various skill levels and ranks cooperating with each other and interacting dynamically, allowing unintended events to occur frequently [18–20]. For better team communication, it is ideal that all the OR team members perform their roles and tasks with a continual understanding of the surrounding dynamic situations. To improve patient safety, all the OR team members should reconsider current decisions and reverify the surgical procedures when a significant discrepancy is observed among their decision-making processes.

In our research, we model how the OR team members understand the situation through **intent inferencing**, where a person's intent is defined as a combination of goals and supporting actions and plans, and is inferred based

[†] Sentinel event: an unexpected occurrence involving death or serious physical or psychological injury, or any process variation for which a recurrence would carry a significant chance of a serious adverse outcome (http://www.jointcommission.org/sentinelevents/).

on probabilistic reasoning. The **team intent** is derived from the intent of individual care team members. The **gap** value is then computed by comparing the likelihoods of possible situations in each member's intent inferencing. For example, a situation involving all care team members and having a high gap value can be interpreted as a medical situation highly vulnerable to medical errors. A person's intent is shaped by his or her perceptions, knowledge, experience, and awareness of environment, just to name a few factors. When each person's intent is embodied by his or her understanding of other team members, the information available may be incomplete and/or inaccurate and should be addressed appropriately when modeling individual reasoning processes. **Bayesian knowledge bases** (BKBs) form the basis for modeling and simulating the OR team members' decision making [21]. By integrating the intentions and beliefs inferred from individual decision-making processes, we identify the discrepancy between intentions and beliefs among the OR team members and use it as an indicator to detect potential medical errors. Although modeling and simulating individual reasoning is a complex and challenging task, by employing the formalism of BKBs, we can handle issues of uncertainty and incompleteness as well as reduce the computational costs required in the **reasoning processes**.

We begin our discussion with a section on *related work*, providing some fundamental background for our research. In our discussion on *team performance*, we introduce our gap analysis procedure and how it can be applied in our domain. Then we provide our current cognitive framework for *surgical intent modeling* and its theoretical foundation of BKBs. Next, in our description of an *empirical study*, we present some real-world medical cases containing errors and provide empirical results for validation. Finally, we provide our conclusion and directions for future research.

RELATED WORK

Communication breakdown has continued to be an issue in medical practice. Although a considerable amount of literature has been published to address the issue, we focus on research devoted to improving communication among medical team members. We classify the major research into three categories: training, checklisting, and intent inferencing.

Training medical care members to enhance patient safety has a long history of research and implementation. In a paper by Awad et al., a special training session, which was based on crew resource management principles, was offered to surgical teams, and the impact of this training was examined by a communication survey collected over several months [22]. The study focused on training OR team members to brief a case before the operation, for the purpose of improving communication. The results of this implementation have been investigated in dedicated hospitals and have shown significant improvement of medical care members' awareness and understanding of the procedures to be performed. Furthermore, the complexity and the dynamics of the OR have been known to parallel those of the aviation environment. Many medical team training systems have employed the principles of aviation crew maneuver training and have shown meaningful improvement [23].

Checklisting is another methodology adapted from aviation crew training principles used to reduce medical errors. The key idea behind the checklist is to standardize processes and to aid the memory of the OR team members. Since this implementation has shown a strong tendency to reduce medical errors, it *is currently in very common use* in the OR. Among a number of variations in checklist design, the two most popular forms are the *to-do list* and the *challenge–verification–response*. The *to-do list* contributes as a systematic way of performing medical procedures, while the *challenge–verification–response* serves as a tool to enhance communication among those involved in the same procedure, such that one party initiates some items from the checklist while another party completes the items [24]. Despite the apparent benefits of the checklist, some medical errors still occur and result in catastrophic outcomes. These causes stem from various sources, many of which are related to poor physician compliance. Some medical care members recite the procedure from memory, not from the checklist; they skip reading the checklist, which would have verified the other party's completeness; some essential items are not included in the checklist, and so on.

Intent inferencing is one of the most advanced techniques dedicated to promoting patient safety because it employs the reasoning tools from artificial intelligence. The research activities include many types of team cooperative tasks, such as central control rooms of power plants, cockpits in aircraft, and medical care members in surgical rooms. In a study by Kanno et al., a two-person team operating a plant control system was simulated by detecting conflict within the team members' intentions [25]. Individual intention was inferred by applying keyhole plan recognition, which searches for a combination of individual mental components with given observables. In addition, there have been multiple studies to maintain quality care by applying computational reasoning and planning, such as ABVAB [26], SPHINX [27], and TraumAID [28]. Our study is in the same line of research. However, we integrate gap analysis to identify potential risks of medical errors and enhance reasoning processes with intent inferencing.

TEAM PERFORMANCE

In the OR, surgery is delivered by several medical professionals, including surgeons, anesthesiologists, and nurses. Medical procedures performed in the

OR are vulnerable to medical errors due to the complexity of surrounding circumstances, where a wide variety of people, medical equipment, activities, and events interact dynamically. Although communication is critical to promote patient safety, it can easily fall apart among medical professionals working as a team in the OR. In this section we address team intent and how the gap is obtained and interpreted in our research to increase situational understanding of team members.

Team Intent

A team is a group of individuals working to achieve common goals. As individual intent leads to a course of actions, team intent leads to the collective actions of team members to achieve common goals. In addition, when team members are better aware of other members' intentions, the team intent can be accomplished in a more effective and efficient manner. High-quality patient care, supposedly a common goal among individual team members, can be better accomplished by enhancing the team intent, which is defined as the collective intent of all team members. However, the intents of the individual team member are not always in accord, often resulting in medical errors. Despite the consensus on the significance of team intent, only a few studies to date have been conducted to address this issue [25]. While existing research concentrates on team intent inference based on a concept of "we-intention," a word coined by Tuomela to represent a set of individual intentions and mutual belief, to simulate a plant controller operated by a two-person team, we focus on quantifying and measuring the level of conflict in teams composed of more than two persons in order to improve patient safety in medical practice [29]. To realize this, we derive team intent from individual intent models and propose a method to quantify team intent by comparing individual intent models. In our understanding, the team intent is time dependent and can either deteriorate or develop, depending on the status of the team members. For example, the team intent may deteriorate when the members of the team perform incorrect procedures or are distracted for personal reasons, such as fatigue or a tragic experience. Analyzing the discrepancy among individuals is a critical step in improving both team members' situational awareness and the team's potential in performing medical procedures and in improving patient safety.

Medical errors are often attributed to the medical care members, especially when they misunderstand the patient, their co-workers, or the surrounding medical situations. For example, a wrong dose of medication may be administered when a nurse misunderstands a doctor's order or the patient's condition. A wrong-site operation can occur when a surgeon is disoriented anatomically by medical images (e.g., MRI or CT) or the nurse prepares the wrong site for operation (e.g., by mistakenly confirming the patient's tattoo as a surgical marking [30]). A retained foreign body occurs when medical care team members leave a piece of operating equipment inside a patient's body when closing an incision. Although anyone can make a mistake, co-workers have the potential to monitor and fix the mistake. Therefore, we would anticipate that a team with more care members would have a better chance of avoiding medical errors when individual team members know their own responsibilities as well as those of the other team members [23]. However, communication easily breaks down in practice, due to the care members' incomplete and inaccurate understanding of their surrounding situations. Therefore, to improve patient care it is essential to enhance the medical care members' understanding and awareness of their environment.

Gaps Among OR Team Members

A medical situation is composed of medical care members and all necessary equipment. Medical errors occur when these components do not function appropriately. A gap is associated with a situation instantiated by these complex components. A large gap suggests that the corresponding situation is vulnerable to medical errors due to discrepancies in individual intents associated with the team. In our research, we consider possible situations from the perspectives of both the team and the individual. To analyze gaps among medical care members, we compute a gap value for a team in a certain situation and use it as a safety measure of the team while performing medical procedures:

$$g(x) = \sum_{i=1}^{n} \sum_{\substack{j=1, \\ i \neq j}}^{n} \text{diff}(\text{ind}(i) - \text{ind}(j)) \tag{9.1}$$

where $g(x)$ represents the gap value of team x in an arbitrary situation composed by n individuals, and $\text{ind}(i)$ denotes the world of an individual i in the same situation. The world of an individual means a common situation as interpreted by that person. Therefore, we can quantify the level of consensus among all members of team x. The gap value computed can be interpreted in various ways, depending on how the individual world is described. With respect to surgical intent modeling, we interpret the gap value as a relative measure of discrepancy among team members in performing medical procedures. Since our intent model is based on probability theory, all individual worlds are represented as the likelihoods of combinations of random variables, and the gap value is obtained by summing the difference of all joint probabilities obtained from individual intent inferencing. Due to the nature of probability theory, comparing the likelihood of different team intent models may not be meaningful in general. However, we can assume that team x is safer than team y in performing a surgical procedure if $gap(x) < gap(y)$, as long as both teams are composed of the same members. Thus, the key difference between the two teams is the team members' intent in performing the medical procedure. For example, when two surgeons cooperate to perform a medical procedure, they need to coordinate with each other when performing their medical actions. Although their actions are different, each has his or her own expectations (beliefs) of the other surgeon. This is also true for other medical care members, such as the anesthesiologist and the nurse, when more care members are involved in performing the same medical procedure. If there is a gap between their intentions and beliefs, this may indicate a potential risk of errors. This may be caused by some care members' lack of experience and knowledge, the distractions they may experience due to fatigue, or the complexity of the procedure.

SURGICAL INTENT INFERENCING

The individual's intent is a psychological concept and can be understood in various ways [31]. In our work, a surgeon's intent is inferred from his or her course of actions and perceptions of the environment. To make this feasible, we need a computational methodology to appropriately represent each person's knowledge and perceptions. We employed Bayesian knowledge bases, a probabilistic knowledge representation, to represent information available in the OR, which is frequently incomplete and uncertain (e.g., information in trauma cases, such as allergies, preexisting conditions, and family history, can often be incomplete in nature because these types of cases emerge rapidly) [32]. Through belief revision with BKBs, we simulate the reasoning processes of health care professionals. In this section we review the basic theory and reasoning processes of BKBs to derive our approach to surgical intent inferencing.

Bayesian Knowledge Bases

Bayesian knowledge bases (BKBs) subsume bayesian networks (BNs) and are represented through directed graphs embracing the causal relationship between pieces of knowledge [33]. Similar to BNs, both graph and probability theories form the theoretical basis for BKBs. A directed graph representation provides a formal and visual expression of causality among pieces of knowledge enclosed, while probability theory guarantees the semantic soundness in decision making under uncertainty and inaccuracy [34]. However, unlike BNs, the BKBs consider partial independence among knowledge pieces and

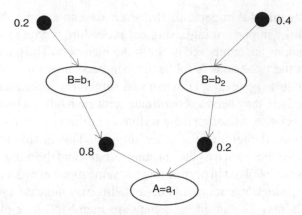

FIGURE 9.1   BKB fragment.

are capable of integrating incompleteness and uncertainty in decision making [32,35]. The nodes of a BKB represent states of random variables, while the arcs denote the causal relationships among the random variables. In particular, the nodes are classified into two types: I-nodes and S-nodes. In an example of small BKBs, as shown in Figure 9.1, the I-nodes, white ovals in the figure, store knowledge to be represented with two random variables, A and B. The dependencies between I-nodes are encoded by S-nodes, while the attached conditional probability indicates the likelihood of the child I-node occurring given that a parent I-node is observed. Consequently, a piece of the knowledge enclosed in the BKB, which can be phrased as $A = a_1$, will occur with an 80% chance when $B = b_1$. In this simple and expressive manner, knowledge can be represented through BKBs.

In general, BNs require a separate conditional probability table containing all possible states of connected random variables, while BKBs focus on partial independence among knowledge. Therefore, BKBs do not require complete knowledge and are capable of reducing complexity when interpreting the knowledge under consideration [36]. Reasoning in BKBs can be implemented in two ways: belief updating and belief revision. Both are based on the dependencies among pieces of knowledge contained, evidence observed prior to the reasoning, and the chain rule:

$$P(X_1, X_2, X_3, \ldots, X_n) = \prod_i^n P(X_i | \text{parents}(X_i)) \tag{9.2}$$

In *belief updating*, the focus is on updating knowledge by computing the posterior probability of any single I-node using Bayes' theorem. In *belief revision*, the most probable world of random variables is derived by computing and comparing joint probabilities of all possible worlds of random variables. Therefore, we can obtain alternative explanations through belief revision [37].

Algorithms performing these BKB reasoning processes have been discussed in detail [36,38]. Consequently, the probabilities of the world obtained in belief revision may become extremely small as the amount of knowledge contained increases. This needs to be interpreted reasonably since the most probable world obtained is one possible explanation that best supports the evidence provided. Naturally, this is only valid regarding the knowledge under consideration. To integrate and aggregate massive knowledge pieces, a fusion algorithm has been developed [39].

Intent Inferencing

Research on intent inferencing has been studied over several decades to represent and understand human decision-making processes and behaviors. *Intent* is an explanation of people's actions and is defined as a combination of the goals that are being pursued, the support for the goals, and the plans to achieve them [40]. To represent human intent through computations, we have designed a system that contains these components and is capable of reasoning through them. Previously, it was applied to adversary intent inferencing on the battlefield, and now we incorporate them for surgical intent inferencing [41–43]. Similar to the adversarial intent inferencing, we integrate components of intent into the structure of BKBs [41]. The knowledge relevant to human intent is categorized into four types: axioms, beliefs, goals, and actions. *Axioms* denote a person's knowledge about himself or herself, whereas *beliefs* denote a person's knowledge about others (including the surrounding environments). *Goals* are used to represent the results that a person wants to achieve. People's actions to be taken to achieve their goals are encoded through *actions*. Figure 9.2 shows these four components arranged in a hierarchical structure,

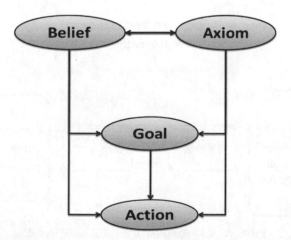

FIGURE 9.2   Hierarchy of interaction between four types of nodes in intent models.

which is recommended to organize correctly and to categorize knowledge when designing BKBs for intent inferencing.

An intent model is composed of a person's knowledge about himself and others and is based on his observations and perceptions. Naturally, the person's knowledge may not be consistent with that of others or even with the world as it is. Therefore, when a group of intent models are collected to compose the team intent, the discrepancies among them are somewhat natural. However, if these individuals undertake their roles under a certain common goal, the discrepancies can be significant, such as the aforementioned potential risk of errors in medical practices.

Surgical Decision Making

According to patient symptoms, the surgeon diagnoses the patient disease. In addition, the patient's vital condition is involved in the surgeon's decision making. The surgeon's decision making is modeled through five major components, as shown in Figure 9.3. The surgeon diagnoses and determines a potential medical procedure based on patient condition, history, and profile. Based on their personal competence, surgeons confirm a procedure to be taken and then determine a course of action to fulfill the objectives of that procedure.

As shown in Figure 9.4, the information associated with the patient's disease, the patient's surgical history, and the patient's family history or genetic information can be included while building a BKB for an individual surgeon. For example, (B)Condition_65105 is inferred from patient's vital signs such as pulse rate, respiration rate, and body temperature. When the patient has vital signs within the normal range, a surgeon can choose an

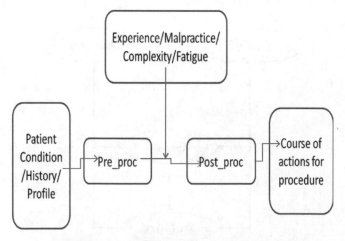

FIGURE 9.3 Skeleton of surgical intent model.

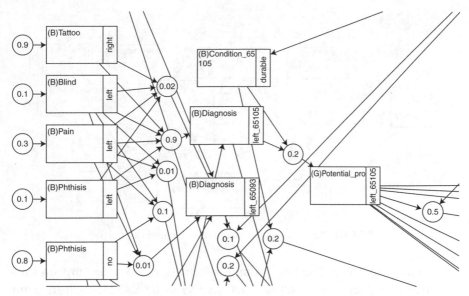

FIGURE 9.4   Patient condition/history/profile.

appropriate procedure. However, if the patient has vital signs that are outside the normal range, a surgeon needs to take an alternative approach for ensuring patient safety. Therefore, it is recommended that an alternative procedure with lower complexity or risk in modeling an individual surgeon's decision making be addressed. Although a surgeon prefers a certain procedure due to her own specialty in general, it is highly recommended that an alternative approach be consided for patient safety, especially for a patient with comorbidities. By linking alternative procedures systematically, an individual intent model can assist a real surgeon effectively since it is possible to build a model to hold more knowledge than a real surgeon can have. In Figure 9.4, two alternative procedures, such as enucleation (65105) and evisceration (65093), are considered [44].

Figure 9.5 shows how personal competence is inferred in the surgeon's decision-making process. This component is composed of various contributing factors: *experience* associated with the procedure, *malpractice* in the past, *complexity* of the procedure, and *fatigue*. These factors are embedded in an individual surgeon's BKB to simulate how a surgeon's decision can be changed by these factors. The information can be derived from the quality of medical school, the postgraduate medical training period, the distribution of procedures the surgeon has performed previously, and the surgeon's recent daily schedule. After a care member predetermines his medical procedure (i.e., pre_proc in Figure 9.3) to perform based on the patient information, he may keep his previous decision or change the procedure, depending on his

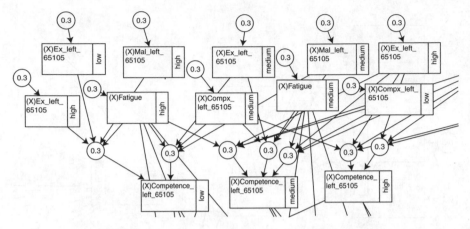

FIGURE 9.5   Experience/malpractice/complexity/fatigue.

personal competence. The background information about the surgeon himself is encoded as *axioms*, represented as (X) in BKBs. Through inferring
personal competence from the background information, a potential risk when
a highly experienced surgeon becomes very fatigued can be captured. For example, a surgeon with high experience and low malpractice can have a lower
("medium") competence level when she becomes highly fatigued than when
she is not fatigued at all. Other OR team members' personal competences are
inferred in the same way, but can be specified differently, depending on their
specialties in performing medical procedures.

Once the surgeon confirms the procedure to be performed, a course of
action is determined. The order of actions can be important, although some
actions are reversible [45]. Many patients in the OR have life-threatening injuries or illnesses, and it is important that a surgeon takes the correct action at
the appropriate time to ensure patient safety. The previous actions completed
by the surgeon and by the other team members need to be considered to
correctly determine the next action to be taken. To simulate dynamically, the
status of the OR should be monitored continuously, and the information obtained needs to be placed into consideration. According to all the information
provided, the most probable action to be taken can be predicted through belief
revision of BKBs. Figure 9.6 shows a part of an ophthalmologist's BKB built
for our simulation, representing a *course of actions for procedure*. This part
can be delineated differently depending on other OR team members' roles
and tasks in performing medical procedures as a team.

EMPIRICAL STUDY

To demonstrate the practical aspects of our surgical intent inferencing, we
built BKBs representing medical professionals and conducted experiments

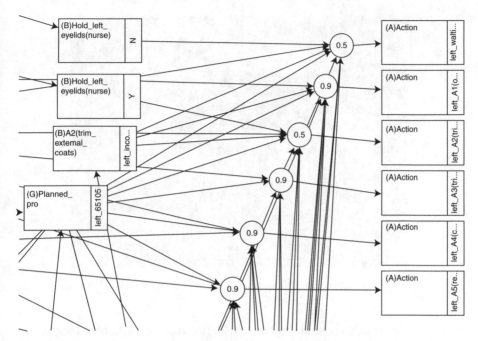

FIGURE 9.6   Course of actions for procedure.

to validate their modeling capability. Previously, we simulated a "hand-off" case of a woman having a breast pain as an example of an adverse event [21]. Communication breakdown between general and plastic surgeons as it pertained to patient hand-off was clearly identified through our approach of surgical intent inferencing. In this paper we focus, however, on team performance by considering three medical professionals: an ophthalmologist, an anesthesiologist, and an OR nurse. In doing so, we computed several gap values from different possible team situations and identified in which situation the team is more prone to risk than others. This experimental study is based on the case published by B. Jericho at the Illinois Medical Center [30]. Although the devastating event was prevented after the surgeon reverified the surgical site in the original case report, we explored hypothetical scenarios in order to address other potential sources of errors based on the case.

Case Study

An 18-year-old male with a history of tobacco, alcohol, and substance abuse came to the hospital with blindness caused by a gunshot wound at the eye that he sustained five months ago. His injured left globe caused the blindness. His left eye became blind, painful, and phthisical (involuted). He was scheduled for two consecutive procedures under general anesthesia: left enucleation

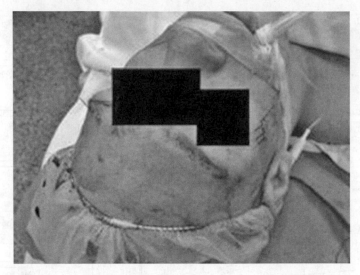

FIGURE 9.7   Tattoo with "ILL" in right eye and surgical mark in left eye [4].

with placement of an orbital implant, and left suture tarsorrhaphy. Despite the patient's ophthalmologic problem, his vital signs were all normal. In addition, he had a dark blue tattoo near his right eye with the initials "ILL". In order to indicate the correct site for the surgery, the left eye was appropriately marked as shown in Figure 9.7.

A preoperative nurse inspecting the care of the patient initially took the tattoo of initials as the surgical site marking and prepared the right eye for surgery. Although there was a chance to reverify the correct site of the operation as indicated on the consent form, the OR nurse confirmed that the tattoo of initials near the right eye was the surgical site marking and proceeded with the surgery without checking against the surgical consent form. Both the anesthesiologist and the ophthalmologist performed the operation on the right eye of the patient, and the medical mistake was discovered when the patient recovered several hours later. To prevent this devastating event, the surgeon should have clarified the correct site of surgery before the operation by examining the patient's consent form, medical history, and condition. If any of the medical care members had reverified the correct site of the operation and covered the tattoo with opaque tape to avoid further confusion in the OR prior to the surgery, this adverse event could have been prevented.

Simulation Results

We built BKBs for three medical care members associated with the case: the ophthalmologist, the anesthesiologist, and the OR nurse. The BKBs, were

TABLE 9.1   Size of BKBs

	RVS.	I-no.	CONN.	S-no. (Rules)	Cond.
Ophthal.	32	112	4.02	149	2.02
Anesthesia	13	33	4.88	60	1.68
OR nurse	21	52	4.48	82	1.84

constructed based on the behavioral patterns and the perceptions of the care members in the OR. Table 9.1 shows the overall scale of BKBs, including the number of random variables (RVS.), the number of I-nodes (I-no.), the average connectivity (CONN.), the number of S-nodes (rules) (S-no.), and the average number of conditions for each rule contained (Cond.) in the BKBs. Through belief revision with the BKBs, each care member's intent was inferred by computing the most probable world composed of random variables under consideration. We assumed that all care team members' intentions and beliefs are identified while they determine the procedures or actions to be taken.

While simulating the case, we considered the vital signs with body temperature, pulse rate, and respiration rate. When the body temperature is within the range 97.8 to 99°F, the pulse rate is within 60 to 100 beats per minute, and the respiration rate is within 15 to 20 breaths per minute, we assume that the patient is sustainable for surgery. For surgical procedures, we considered two types of eye surgeries: enucleation (65105) and evisceration (65093). While performing evisceration, the surgeon removes the ocular contents but preserves sclera and sometimes the cornea. While performing enucleation, the surgeon removes the entire globe and a portion of anterior optic nerve. For some patients who are not enstainable for any type of enucleation, we consider evisceration (65093) as an alternative with less anatomical disruption. To speculate the site of operation associated with medical errors in the experiments, we separate each procedure into two parts: in this case, L65105 and R65105.

Surgeon (Ophthalmologist's) BKB   An ophthalmologist diagnoses a patient based on his condition, such as blindness, pain, and phthisis (involuted). When the patient has all the ophthalmological problems in his left eye, the ophthalmologist considers the vital sign of the patient before he decides a potential procedure to be taken. As shown in Table 9.2, the surgeon diagnoses enucleation (65105) for the patient's left side with a high probability, but he may also choose evisceration (65093) without serious consideration of the wound even if it is less likely. It is possible for him to plan enucleation on the right side by mistake with a low probability. We validated the surgeon's diagnosis with the evidence of the right side as well, as shown in Table 9.2.

TABLE 9.2   Surgeon's Diagnosis

Evidence				Target Variables	Probability
(B)Blind	(B)Pain	(B)Phthisis	(B)Tattoo	(B)Diagnosis	
left	left	left	right	L65105	2.91655e-08
				L65093	8.74964e-09
				R65105	3.64569e-09
right	right	right	left	R65105	2.91655e-08
				R65093	8.74964e-09
				L65105	3.64569e-09

In addition to the patient's symptoms relevant to the ophthalmological problems, the surgeon considers the vital signs of a patient when determining the procedure to be taken. When all three major vital signs are within the normal range (which are represented by (B)Pulse=normal, (B)Respiration=normal, and (B)Body=normal in the BKB), a surgeon can consider the patient durable for both procedures with the highest probability, which is denoted by "Y" in Table 9.3.

When the patient has a high or low body temperature while his other vital signs are normal, a surgeon can consider the patient durable for evisceration (65093) but not for enucleation (65105). If any other vital sign of a patient is out of the normal range, the surgeon considers the patient not durable for either of the two procedures with the highest probability. We investigated how a patient's durability is inferred from the patient's vital signs and obtained the results we expected, as shown in Table 9.3. We omitted cases in which the surgeon determines that the patient is not durable for either procedure under consideration, since they are beyond our research focus.

In addition to the aforementioned patient information, a surgeon determines a procedure based on his personal preference. We implemented a surgeon's

TABLE 9.3   Vital Signs

Evidence			Target Variables	
(B)Pulse	(B)Respiration	(B)Body	(B)Condition_65105	(B)Condition_65093
low	low	low	N	N
normal	normal	low	N	Y
normal	normal	normal	Y	Y
normal	normal	high	N	Y
high	high	high	N	N

TABLE 9.4   Surgeon's Competence Inferencing

Evidence				Target Variables
(X)Experience	(X)Malpractice	(X)Complexity	(X)Fatigue	(X)Competence
low	high	high	high	low
medium	high	high	high	medium
medium	medium	medium	medium	medium
medium	medium	medium	high	medium
high	low	low	low	high
high	low	low	medium	medium
high	low	low	high	low
high	low	medium	low	medium

personal preference in choosing a medical procedure with (X)Competence, as shown in Table 9.4. In particular, we included four factors (experience, malpractice, complexity, and fatigue) for implementing the component for inferencing a surgeon's personal competence. We assume an increasing personal competence level as the level of experience increases, or as the level of malpractice, complexity, or fatigue decreases. For example, when a surgeon has a level of low experience, high malpractice, high complexity, and high fatigue, her competence level becomes low. However, if her level of experience increases to the level of medium, her competence level advances to medium when other factors remain the same. Among all possible states, we provide only eight interesting states here to explain how each of these factors influences a surgeon's personal competence. Although a surgeon's personal competence decreases with increasing level of fatigue in general, the competence level remains the same sometimes despite a change in the surgeon's fatigue level. For example, when a surgeon has a level of low experience, high malpractice, and high complexity, her level of personal competence is assumed as low with the highest probability, regardless of her fatigue level, since her best personal competence level is low and can barely be improved or worsened with the change of her fatigue level. Through this design, we made the level of experience more influential than other factors, although it is not a dominant factor.

We assume that a surgeon's competence influences her probability of making a mistake in determining a correct procedure. Therefore, as the level of competence decreases, the error probability increases while the probability of performing correct procedure decreases. As shown in Table 9.5, we validated the ophthalmologist's BKB with a set of evidence including three different levels of personal competence.

TABLE 9.5   Influence of Personal Competence

		Evidence			Target Variables	Probability
(B)Blind	(B)Pain	(B)Phthisis	(B)Tattoo	(X)Competence	(B)proc	
left	left	left	right	high	L65105	2.91655e-08
					L65093	2.18741e-09
					R65105	3.64569e-10
left	left	left	right	medium	L65105	2.55198e-08
					L65093	4.37482e-09
					R65105	1.82284e-09
left	left	left	right	low	L65105	1.82284e-08
					L65093	8.74964e-09
					R65105	3.64569e-09

Based on the medical situations involved, a surgeon's next action is determined. Table 9.6 represents a surgeon's next action, as predicted with the corresponding evidence set. With respect to enucleation (65105) procedure, we considered a course of action as follows:

A1. Open and order the OR nurse to hold the eyelids.

A2. Trim away the external coats of the eye from the eyeball.

A3. Trim away the extraocular muscles from the eye surface.

A4. Cut the optic nerve.

A5. Remove the entire eyeball.

A6. Put an orbital implant into the socket.

A7. Close the tissues and do tarsorrhaphy.

TABLE 9.6   Next Action Prediction

			Evidence					Target Variables
(B)A1	(B)A2	(B)A3	(B)A4	(B)A5	(B)A6	(B)A7	(B)Nurse	(A)Action
N	N	N	N	N	N	N	Na	A1
Y	N	N	N	N	N	N	N	Waiting
Y	N	N	N	N	N	N	Y	A2
Y	Y	N	N	N	N	N	Na	A3
Y	Y	Y	N	N	N	N	Na	A4
Y	Y	Y	Y	N	N	N	Na	A5
Y	Y	Y	Y	Y	N	N	Na	A6
Y	Y	Y	Y	Y	N	N	Na	A7

When none of the actions are completed, a surgeon is supposed to take action A1. After the surgeon completes action A1 [represented as $(B)A1 = Y$ in Table 9.6], the surgeon's next action depends on his belief about the nurse's status. If the nurse is ready to hold the patient's eyelid, the surgeon performs the next action (A2). Otherwise, the surgeon waits for the nurse to be ready to assist. For his subsequent actions, the surgeon does not rely on his belief on the nurse, but rather, determines his next action depending on the completeness of his prior actions.

Anesthesiologist's BKB   The duty of the anesthesiologist consists mainly of selecting anesthesia prior to the surgery and adjusting anesthesia if necessary during the surgery. Prior to a surgery, the anesthesiologist interviews the patient to form a detailed plan about anesthesia injection and to learn about precautions that must be addressed. During the surgery, the anesthesiologist monitors the patient's vital signs and acts according to the patient's physical state. Similar to the surgeon's BKB, the anesthesiologist's personal competence is inferred from his level of experience and fatigue, as shown in Table 9.7.

With regard to medical errors, we assume that a highly competent anesthesiologist can better select and administer anesthesia than an anesthesiologist of low competentcy. To validate the hypothesis, we varied the personal competence of the anesthesiologist and observed how the chance of making a mistake in determining anesthesia was changed. As we expected, the results obtained show that the probability that the anesthesiologist will select and administer an appropriate anesthesia increases, while the probability of his making mistakes decreases with increasing personal competence, as shown in Table 9.8.

In addition to the personal competence and its impact on the probability of medical errors, we addressed an anesthesiologist's decision making in selecting an appropriate anesthesia for the patient. In our current model, we considered diabetes and cardiovascular and pulmonary problems as shown

TABLE 9.7   Inference of Personal Competence

Evidence		Target Variables
(X)Experience	(X)Fatigue	(X)Competence
low	high	low
low	low	medium
high	low	high
high	high	medium

TABLE 9.8   Influence of Personal Competence

Evidence					Target Variables	Probability
(B)Blind	(B)Pain	(B)Phthisis	(B)Tattoo	(X)Com	(G)Anesthesia	
left	left	left	right	high	General	1.64025e-06
					Left_local	3.2805e-07
left	left	left	right	medium	General	1.47622e-06
					Left_local	4.92075e-07
left	left	left	right	low	General	1.3122e-06
					Left_local	6.561e-07

TABLE 9.9   Validation for Selecting Anesthesia

Evidence			Target Variables
(B)Diabetes	(B)Cardiovascular	(B)Pulmonary	(B)General_refused
N	N	N	N
Y	N	N	Y
N	Y	N	Y
N	N	Y	Y

in Table 9.9, since these are critical conditions leading to fatal consequences under general anesthesia. When the patient has any of these critical diseases, the anesthesiologist needs to use local anesthesia, even if general anesthesia is preferred for eye surgeries such as enucleation or evisceration. We varied the patient's condition and confirmed that the results obtained are consistent with expectation, as shown in Table 9.9. The target variable, (B)General_refused=Y, denotes the patient's condition, which is too risky for general anesthesia.

We assumed the course of actions for the anesthesiologist as "choose_drug→injection→infiltration→monitoring." To implement this, we introduced four random variables: (B)Choose_drug, (B)Injection, (B)Infiltration, and (B)Monitoring, and represent different status levels of the OR by instantiating these random variables differently. As shown in Table 9.10, the anesthesiologist's BKB correctly predicted the next action to take, as we expected.

OR Nurse's BKB   To validate that the nurse's model is a true representation of a real nurse's decision making, we focused on how the nurse's personal competence is inferred from his experience and fatigue and how the nurse determines the correct action to take in any given situation. First, the inference

TABLE 9.10 **Validation of Actions**

	Evidence			Target Variable
(B)Choose_drug	(B)Injection	(B)Infiltration	(B)Monitoring	(A)Action
N	N	N	N	Choose_drug
Y	N	N	N	Injection
Y	Y	N	N	Infiltration
Y	Y	Y	N	Monitoring

of the nurse's personal competence was tested. We assumed that the nurse's personal competence increases as the nurse's level of experience increases or the level of fatigue decreases. The states of target variables in Table 9.11 show the results obtained, which are consistent with our assumption.

We also assume that the chance of making a mistake depends on the nurse's personal competence. The OR nurse can make two types of mistakes: He can misunderstand the surgeon's order, or he can misunderstand the patient's condition. Occasionally, OR nurses assume a different procedure to be determined by the surgeon, due to his lack of knowledge or experience. Since the nurse's actions depend heavily on the procedure to be performed by the surgeon, his misunderstanding of the surgeon's intent can lead to an adverse outcome. With increasing personal competence, the chance that the nurse understands other team members correctly increases, while the chance that the nurse makes mistakes decreases. Table 9.12 validates how the probability changes with the varying personal competence level of the OR nurse.

We assumed that the nurse performs three types of actions: checking the patient's vital signs, waiting for the order, and following the order, which is "hold_eyelid" in this case. As shown in Table 9.13, the nurse waits for the surgeon to give an order when there is no urgency (i.e., the patient's vital signs are within the normal range) or for a specific order to take. When the surgeon orders a certain action, the OR nurse follows the order. When the patient's vital signs move dramatically [i.e., (B)Vital_request=Y], the OR

TABLE 9.11 **Inference of Personal Competence**

	Evidence	Target Variables
(X)Experience	(X)Fatigue	(X)Competence
low	high	low
low	low	medium
high	low	high
high	high	medium

TABLE 9.12   Influence of Personal Competence

Evidence					Target Variables	Probability
(B)Blind	(B)Pain	(B)Phthisis	(B)Tattoo	(X)Competence	(G)Nast	
left	left	left	right	high	L65105	1.77147e-06
					R65105	3.54294e-07
					L65093	5.31441e-07
				medium	L65105	1.41718e-06
					R65105	4.25153e-07
					L65093	5.6687e-07
				low	L65105	1.06288e-06
					R65105	4.60582e-07
					L65093	6.023e-07

TABLE 9.13   Validation with the Planned Procedure of Nurse

Evidence		Target Variable
(B)Order_to_hold	(B)Vital_request	(A)action
N	N	Waiting
Y	N	Hold_eyelid
N	Y	Check_vital
Y	Y	Check_vital

nurse checks the cause of the fluctuation and follows the emergency care routine. Although we assigned a higher priority to emergency care than to the surgeon's order ("hold_eyelid"), these can be hardly separated in practice. We list the experiment setting and results in Table 9.13.

Gap Analysis   We assume that medical errors occur when there is a significant discrepancy among the team members' intent. By comparing individuals' intents, we aim to determine whether or not a team has a high risk. Due to the complexity of the OR and medical processes, the number of possible worlds associated with the case can be tremendous. Even if we consider incomplete and inaccurate worlds of information in our research, the number of possible worlds under consideration is still intractable. To show the applicability of our gap analysis in identifying situations having a high risk of medical errors, in this section we consider a few situations as examples.

S1: No risk of medical error. A patient has an ophthalmological problem in his left eye, and all care members agree on performing enucleation (65105) on the left eye of the patient. With respect to medical errors, this situation is normal and we assume there to be no gap in the team composed of those medical professionals.

S2: Wrong-side preparation. A patient has an ophthalmological problem in his left eye, but the OR nurse makes a mistake while preparing the surgery because of the confusing tattoo near the right eye.

S3: Wrong-side operation. A patient has an ophthalmological problem in his left eye, but the ophthalmologist was disoriented by body symmetry when he read the patient's CT before the operation. He determined to perform enucleation (65105) on the right side.

S4: Misdiagnosis. The ophthalmologist decides to perform evisceration (65093) for the patient without recognizing a severe phthisis on the left eye and expects other care members to work for the same procedure. However, the anesthesiologist and the OR nurse know the patient's condition correctly and expect the ophthalmologist to perform enucleation (65105).

S5: Wrong anesthesia. A patient has an ophthalmological problem in his left eye and the ophthalmologist and the OR nurse prepare the enucleation (65105) on the patient's left side. However, the anesthesiologist decides to use local anesthesia since he was confused by another patient who had diabetes. Even if the patient has a severe phthisis, the anesthesiologist expects other care members to perform evisceration (65093).

As shown in Table 9.14, several medical errors are possible for various reasons, even if the same evidence is given. From the ophthalmologist's perspective, (G)Plan indicates the medical procedure to be taken. From the anesthesiologist's perspective, (G)Plan is to select and administer an appropriate anesthesia. From the OR nurse's perspective, (G)Plan is the procedure in which he assists. In each team member's intent inferencing, his or her belief of other team members was also inferred with the given evidence. For example, the ophthalmologist believes that the anesthesiologist would determine general anesthesia and the OR nurse would assist enucleation on the left side when the ophthalmologist decided to perform enucleation on the left side of

TABLE 9.14   Comparison of Hypothetical Situations

	Description	Ophthalmologist (G)Plan	Anesthesiologist (G)Plan	OR Nurse (G)Plan
S1	Correct	L65105	general	L65105
S2	Wrong-side preparation	L65105	general	R65105
S3	Wrong-side operation	R65105	general	L65105
S4	Misdiagnosis	L65093	general	L65105
S5	Wrong anesthesia	L65105	Left_local	L65105

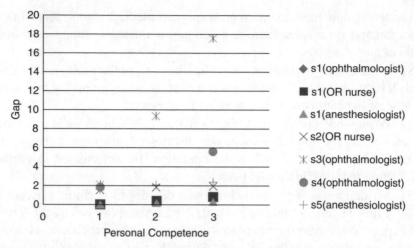

FIGURE 9.8   Team members' personal competences vs. gap.

the patient (i.e., L65105). The shaded cells denote that the medical professionals have different intents and beliefs than other care members, causing a
discrepancy in the team intent. To validate the applicability of gap analysis,
we varied each member's personal competence and computed the gap value
according to equation (9.1). We assume that losing personal competence increases the gap among the team members and leads to a situation riskier than
others. As shown in Figure 9.8, we investigated all five situations with varying team members' personal competence level. Three different levels were
considered for each medical professional. On the x-axis, 1 denotes the highest and 3 the lowest competence level. The y-axis represents the gap value
computed by equation (9.1). For all team members, the gap values computed
increase as the personal competence level is degraded. Considering the situation with respect to a team member's role, we varied the ophthalmologist's
competence level for situations S1, S3, and S4, the anesthesiologist's competence level for situations S1 and S5, and the OR nurse's competence for
situations S1 and S2.

While computing the gap values with the probabilities obtained from belief
revision, we considered the highly competent team member as a baseline in
each situation. Through the experiments conducted with varying personal
competence of all team members, we confirmed that the BKBs' representing
capability was consistent with our expectation.

CONCLUSION

In this study we present a **cognitive computational framework** to simulate
the reasoning processes of medical team members to reduce medical errors

by identifying and resolving gaps among individual care members. Communication breakdown among medical team members has been known to be a major cause of adverse events, and we expect our approach to contribute to diminishing the communication loss and assisting medical care members to better understand the dynamic environments and their co-workers. Among various types of medical errors, thus far we have investigated miscommunication among medical team members caused by misdiagnosis, wrong-site operation, and wrong anesthesia. To accomplish our ultimate research goal, promoting patient safety in the OR, it is necessary to simulate other types of errors, by investigating more test cases.

We consider two future directions: temporal relationships among pieces of information and generalization among various medical procedures. To simulate medical cases dynamically, reasoning and inferencing knowledge with respect to time is essential. Although there is a theory regarding temporal BKBs, the computational complexity hinders its applicability [46]. In addition, general components of surgical intent inferencing need to be formulated, which would be different from the hierarchy of intent inferencing in other domains.

Acknowledgments

This work was supported in part by grant N00014-08-1-0879 from the Office of Naval Research. Earlier preliminary results of this work may be found in Santos et al. [47].

KEY TERMS

intent inferencing	checklisting
team intent	reasoning processes
gap	cognitive computational framework
Bayesian knowledge bases	

REFERENCES

[1] Kohn LT, Corrigan JM. *To Err Is Human.* Washington, DC: National Academy Press, 2000.

[2] Hurwitz B, Sheikh A. *Health Care Errors and Patient Safety.* London: BMJ Books, 2009.

[3] Dalton GD, Samaropoulos XF, Dalton AC. Improvements in the safety of patient care can help end the medical malpractice crisis in the United States. *Health Policy*, 86(2–3):153–162, May 2008.

[4] Pronovost PJ, Thompson DA, Holzmueller CG, et al. Toward learning from patient safety reporting systems. *J Crit Care*, 21:305–315, 2006.

[5] Kovacs G, Croskerry P. Clinical decision making: an emergency medicine perspective. *Acad Emerg Med*, 6(9):947–952, 1999.

[6] McIntyre R, Stiegmann GV, Eiseman B. *Surgical Decison Making*. Philadelphia: W.B. Saunders, 2004.

[7] Horwitz LI, Krumholz H M, Green ML, Huot SJ. Transfers of patient care between house staff on internal medicine wards: a national survey. *Arch Intern Med*, 166:1173–1177, 2006.

[8] Landro L. Hospitals combat errors at the "hand-off." *The Wall Street Journal*, June 28, 2006.

[9] Eastridge BJ, Jenkins D, Flaherty S, Schiller H, Holcomb JB. Trauma system development in a theater of war: experiences from Operation Iraqi Freedom and Operation Enduring Freedom. *J Trauma*, 61(6):1366–1372, Dec. 2006.

[10] Montgomery SP, Swiecki CW, Shriver CD. The evaluation of casualties from Operation Iraqi Freedom on return to the continental United States from March to June 2003. *J Am Coll Surgeons*, 201(1):7–12, July 2005.

[11] Guise JM. Teamwork in obstetric critical care. *Best Pract Res Clin Obstet Gynaecol*, 22(5):937–951, 2008.

[12] Makary MA, Sexton JB, Freischlag JA. Operating room teamwork among physicians and nurses: teamwork in the eye of the beholder. *J Am Coll Surgeons*, 202(5):746–752, 2006.

[13] Williams KA, Rose WD, Simon R. Teamwork in emergency medical services. *Air Med J*, 18(4):149–153, 1999.

[14] Reader TW, Flin R, Cuthbertson BH. Communication skills and error in the intensive care unit. *Curr Opin Crit Care*, 13:732–736, 2007.

[15] Joint Commission on Accreditation of Healthcare Organizations. *Sentinel Events: Evaluating Cause and Planning Improvement*. Oakbrook Terrace, IL: JCAHO, 1998.

[16] Lingard L, Espin S, Whyte S., et al. Communication failure in the operating room: observational classification of recurrent types and effects. *Qual Saf Healthcare*, 13(5):330–334, 2004.

[17] Parush A, Kramer C, Foster-Hunt T, Momtahan K, Hunter A, Sohmer B. Communication and team situation awareness in the OR: implications for augmentative information display. *J Biomed Inf*, 2010.

[18] Christian CK, Gustafson ML, Roth EM, et al. A prospective study of patient safety in the operating room. *Surgery*, 139(2):159–173, 2006.

[19] Murff HJ, Patel VL, Hripcsak G, Bates DW. Detecting adverse events for patient safety research: a review of current methodologies. *J Biomed Inf*, 36:131–143, 2003.

[20] Alvarez G, Coiera E. Interdisciplinary communication: an uncharted source of medical error? *J Crit Care*, 21:236–242, 2006.

[21] Santos E, Jr., Rosen J, Kim KJ, Yu F, Li D. *Reasoning About Intentions in Complex Organizational Behaviors: Intentions in Surgical Handoffs*. DI2AG Technical Report-2010-100. Hanover, NH: Thayer School of Engineering, Dartmouth College, 2010.

[22] Awad SS, Fagan SP, Bellows C, Albo D, Green-Rashad B, Garza M, Berger DH. Bridging the communication gap in the operating room with medical team training. *Am J Surg*, 190(5):770–774, Nov. 2005.

[23] Helmreich RL, Wilhelm JA, Klinect JR, Merritt AC. Culture, error and crew resource management. In *Improving Teamwork in Organizations: Applications of Resource Management Training*. Hillsdale, NJ: Laurence Erlbaum Associates, 2001.

[24] Gawande, A. *The Checklist Manifesto: How to Get Things Right*. New York: Metropolitan Books, 2009.

[25] Kanno T, Nakata K, Furuta K. A method for conflict detection based on team intention inference. *Interact Comput*, 18(4):747–769, July 2006.

[26] Wong WS, Felix LKS, So YT. The recent development and evaluation of a medical expert system (ABVAB). *Int J Bio-Med Comput*, 25(2–3):223–229, Apr. 1990.

[27] Fieschi M. Towards validation of expert systems as medical decision aids. *Int J Bio-Med Comput*, 26(1–2):93–108, July 1990.

[28] Webber B, Carberry S, Clarke JR, Gertner A, Harvey T, Rymon R, Washington R. Exploiting multiple goals and intentions in decision support for the management of multiple trauma: a review of the TraumAID project. *Artif Intell*, 105(1–2):263–293, Oct. 1998.

[29] Tuomela R, Miller K. We-intentionion. *Philos Stud*, 53:367–389, 1987.

[30] Jericho BG, Campise-Luther R, Changyaleket B, Setabutr P, Sajja K, McDonald, T. Facial tattoo and wrong site surgery: which side are we operating on? *Internet J Anesthesiol*, 21(2): 2009.

[31] Manterea S, Sillinceb JA. Strategic intent as a rhetorical device. *Scand J Manag*, 23:406–423, 2007.

[32] Santos ES, Santos E, Jr. Reasoning with uncertainty in a knowledge-based system. In *Proceedings of the Seventeenth International Symposium on Multiple-Valued Logic*, Boston, MA, 1987, pp. 75–81.

[33] Pearl J. *Probabilistic Reasoning in Intelligent Systems*. San Francisco: Morgan Kaufmann, 1988.

[34] Santos E, Jr., Santos ES, Shimony SE. Implicitly preserving semantics during incremental knowledge base acquisition under uncertainty. *Int J Approx Reason*, 33(1):71–94, 2003.

[35] Santos E, Jr., Santos ES. A framework for building knowledge-bases under uncertainty. *J Exp Theor Artif Intell*, 11:265–286, 1999.

[36] Rosen T, Shimony SE, Santos E, Jr. Reasoning with BKBs: algorithms and complexity. *Ann Math Artif Intell*, 40(3–4):403–425, 2004.

[37] Santos E, Jr. On the generation of alternative explanations with implications for belief revision. In *Proceedings of the Seventh Conference on Uncertainty in Artificial Intelligence*, Los Angeles, 1991. 337–347.

[38] Santos E, Jr. Verification and validation of knowledge-bases under uncertainty. *Data Knowledge Eng*, 37:307–329, 2001.

[39] Santos E, Jr., Wilkinson JT, Santos EE. Bayesian knowledge fusion. In *22nd International FLAIRS Conference*. Sanibel Island, FL: AAAI Press, 2009.

[40] Searle JR. *Intentionality: An Essay in the Philosophy of Mind*. New York: Cambridge University Press, 1983.

[41] Santos E, Jr., Zhao Q. Adversarial models for opponent intent inferencing. In *Adversarial Reasoning: Computational Approaches to Reading the Opponent's Mind*. Kott A, McEneaney W, Eds. Boca Raton, FL: Chapman & Hall/ CRC press, 2006.

[42] Santos E, Jr. A Cognitive Architecture for adversary intent inferencing: knowledge structure and computation. In *Proceedings of the SPIE 17th Annual International Symposium on Aerospace/Defense Sensing and Controls: AeroSense 2003*.

[43] Santos E, Jr., Negri A. Constructing adversarial models for threat/enemy intent predictiong and inferencing. In *Proceedings of the SPIE Defense and Security Symposium*, Orlando, FL, 2004. pp. 77–88.

[44] American Medical Association. *Current Procedural Terminology (CPT)*. Chicago, AMA Press, 2004.

[45] Healey PM, Jacobson EJ. *Common Medical Diagnoses: An Algorithmic Approach*. Philadelphia: W.B. Saunders, 1994.

[46] Pioch NJ, Melhuish J, Seidel A, Santos E, Jr., Li D, Gorniak N. Adversarial intent modeling using embedded simulation and temporal Bayesian knowledge bases. In *SPIE Defense, Security and Sensing*, Orlando, FL, 2009.

[47] Santos E, Jr., Kim KJ, Yu F, Li D, Jacob E, Katona L, Rosen J. A method to improve team performance in the OR through intent inferencing. In *7th EUROSIM Congress on Modeling and Simulation*, Prague, Czech Republic, 2010.

10 Future of Modeling and Simulation in the Medical and Health Sciences

RICHARD M. SATAVA

INTRODUCTION

It has been over 100 years since the last change in medical education. With advances in modeling and simulation moving into health care, there are unique opportunities once again to revolutionize medical education, using simulation, advanced curricula, objective metrics, and validated outcomes to change the current subjective time-based apprentice model to one of objective structure competency-based training and assessment. Entirely new opportunities in objective assessment, team training, preoperative warm-up, and surgical rehearsal provide a new foundation for medical education based on advances in simulation technology, information science, and curricula development. The revolution has just begun.[†]

In 1908, Flexner conducted the first comprehensive review of medical education in the United States, and changed the paradigm from simple apprenticeship to a more structured, formal curriculum-based educational process [1]. Over the intervening century, many iterations and improvements upon this model have been made, although the process remained a subjective assessment and personal reference-based system (except for written and

[†]The opinions or assertions contained herein are the private views of the author and are not to be construed as official, or as reflecting the views of the Departments of the Army, Navy, or Air Force; the Defense Advanced Research Projects Agency; or the Department of Defense. Additionally, this is a declared work of the U.S. government and as such is not subject to copyright protection in the United States.

Modeling and Simulation in the Medical and Health Sciences, First Edition. Edited by John A. Sokolowski and Catherine M. Banks.
© 2011 John Wiley & Sons, Inc. Published 2011 by John Wiley & Sons, Inc.

computer-administered knowledge examinations). However, during this same period, beginning in the late 1920s and early 1930s, the military began exploring opportunities for simulation-based training, especially in the area of flight simulation. By the advent of World War II, Link had begun providing "carnival-ride" level flight simulators; however, with the proper curriculum such simplistic simulators were able to reduce crash landings dramatically in night or bad-weather conditions. The rest is history: Aviation began a complete industry in modeling and simulation that is the envy of the world, used in both military and commercial aviation and producing a remarkable safety level. The military also expanded simulation to encompass literally every aspect of training and assessment. Recent research has extended this to include even the "soft sciences," such as human social, behavioral, and cultural (HSBC) simulation and training based on a new understanding of the neurocognitive sciences and advances in simulation fidelity.

Medical simulation is still lagging woefully behind. Even though the first manikin-based simulator by Gaba and DeAnda [2] was developed in 1988 and the first computer-based virtual reality (VR) surgical simulator was developed by Satava [3] in 1993, it wasn't until 1997, when Reznick et al. [4] and Derossis et al. [5] developed and validated an objective and structured curriculum for training and assessment, that the medical and surgical communities began to take simulation seriously. A key turning point came in 2002 when Seymour et al. [6] validated the first simple computer-based virtual reality simulator in a randomized, controlled, double-blind clinical trial which combined with the growing evidence to encourage the adoption of simulation as a valuable educational tool. The result of these above combined efforts over decades has been the adoption by the **Residency Review Committee** (RRC) of the **American Council of Graduate Medical Education** (ACGME) of the mandate in July 2008 that required every surgical residency training program to have access to simulation training and assessment [7]. The **American College of Surgeons** (ACS) endorsed the Fundamentals of Laparoscopic Surgery (FLS) course of the **Society of American Gastrointestinal and Endoscopic Surgery** (SAGES), and the American Board of Surgery (ABS) followed shortly in 2009 with the requirement that every applicant resident for surgical board examination must have documentation of completion of the SAGES/FLS course included as part of the application process. Failure to include the SAGES/FLS certification resulted in the application being returned as incomplete and the resident not being able to sit for their certification exam. Even though a resident had "completed" training, he or she would not be able to become a board-certified surgeon [8].

A number of other critical changes have occurred in the area of technical skills training (which include both the cognitive and the psychomotor skills) that have contributed to the revolution. Some of the important factors have been the defining of key competencies (the ACGME classification of six competencies), comprehensive and structured curricula, objective metrics for assessment, changing to proficiency-based training and assessment, standardized validation methodologies, and new simulation technologies. These processes are critical because they redefine the way that modeling and simulation will be researched, developed, and adopted in the future. Not adhering to the processes usually results in a simulation that is not accepted for certification by the governing medical authority. We discuss each to emphasize their importance, including a summary of a number of the new (and future) applications for simulation.

FUTURE PROCESSES FOR MODELING AND SIMULATION

Currently, a number of processes are being put into place in an effort to "standardize" the development, validation, and certification of medical modeling and simulation. Awareness of these processes is important in planning the design and execution of a full curriculum. For our purposes here, the term *curriculum* refers to the entire life-cycle process from initial consensus conferences to the final acceptance as a mandate for certification (see Figure 10.1). It is important to note that the entire process is dynamic and that it begins with metrics and ends with feedback from the certifying authorities regarding the metrics, for continual improvement of the training and certification process.

Definition of Competence

The 2003 ACGME totally revised what elements of medical education were required for a person to be considered "competent" and approved by the federation of all the medical specialty boards [American Board of Medical Specialties (ABMS)]. These included:

1. Knowledge
2. Patient care
3. Interpersonal and communication skills
4. Professionalism
5. Systems-based practice
6. Lifelong learning

What is interesting is the fact that only the first two critical components (knowledge and patient care) have been formally taught. The other four had

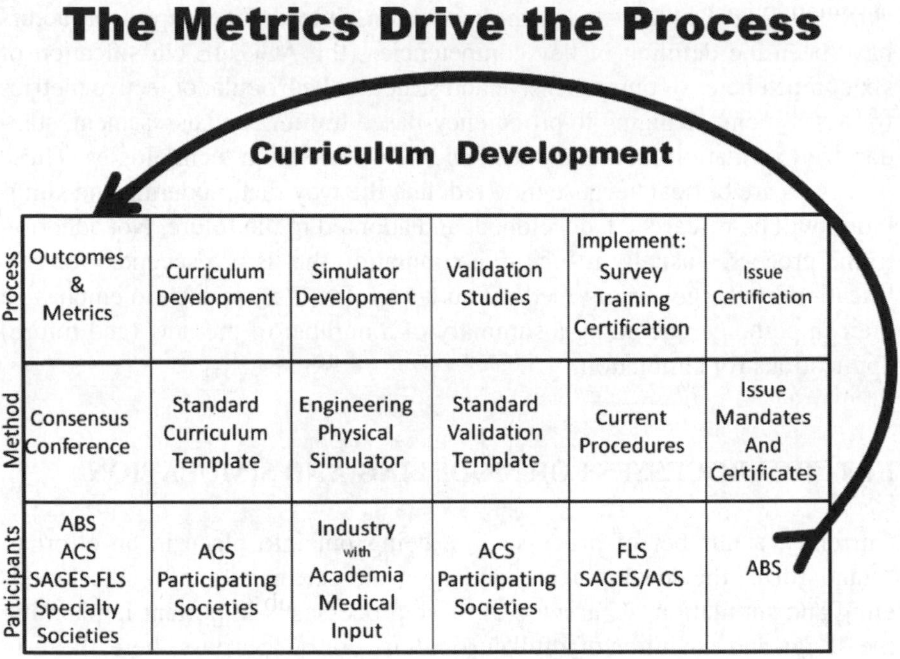

The Metrics Drive the Process

Curriculum Development

	Outcomes & Metrics	Curriculum Development	Simulator Development	Validation Studies	Implement: Survey Training Certification	Issue Certification
Process						
Method	Consensus Conference	Standard Curriculum Template	Engineering Physical Simulator	Standard Validation Template	Current Procedures	Issue Mandates And Certificates
Participants	ABS ACS SAGES-FLS Specialty Societies	ACS Participating Societies	Industry with Academia Medical Input	ACS Participating Societies	FLS SAGES/ACS	ABS

FIGURE 10.1   Life cycle of curriculum development.

been expected to be "learned" through informal mentoring and apprentice methods. There was no formal educational process, no definitions or no curricula, and certainly, no measurements. On the other hand, this vacuum provided a huge opportunity for simulation of all aspects. Early responses included team-based training (modeled after aviation's crew resource management training) and **objective structured clinical examinations** (OSCEs). The team training utilized a manikin simulator for specific anesthesia technical skills as well as communication, professionalism, and systems-based skills, while the OSCE utilized actual "patient actors." Validation demonstrated the value added to such training, and both have been accepted as important clinical training tools.

Objective Structured Curricula and Assessment Tools

Once manikins and patient actors began to be incorporated into a skills training curriculum, it became apparent that a fully structured curriculum approach was needed. It was at this time that the concept evolved that skills training "is not about a simulator, it is about the curriculum." From this need the **objective structured assessment of technical skills** (OSATS) was developed and validated. This was the first time that specific quantifiable metrics were assigned to each simulated task, allowing the faculty observers to accurately (consistently and reliably) assess and grade the performance of a resident's technical skill. It is essential that all outcomes measures, both quantitative and qualitative, be described in absolutely unambiguous terms—to the point that evaluators can all agree as to what is or is not a correct measure, error, or action. This precise description or metric provided a method to train faculty to assess performance with an interrater reliability of $r \geq 0.80$, the standard metric for validation.

In a similar fashion, the OSCE began developing very precise measurements for the behavior and performance of a student who was interviewing or examining a patient (actor), not only for history-taking and physical examination skills but also for communication and professionalism skills. This development resulted in a standardized approach to the curriculum. One example of a template for a curriculum is shown in Figure 10.2. Of importance is the inclusion of errors as part of the didactic curriculum. It was discovered during the validation of a number of the simulators that some subjects were repeatedly making the same mistake.

When asked about the repetitious errors, the subjects' invariably commented that they did not know their actions were errors—no one had instructed them on errors, so they continued to make the same mistakes. Subsequently,

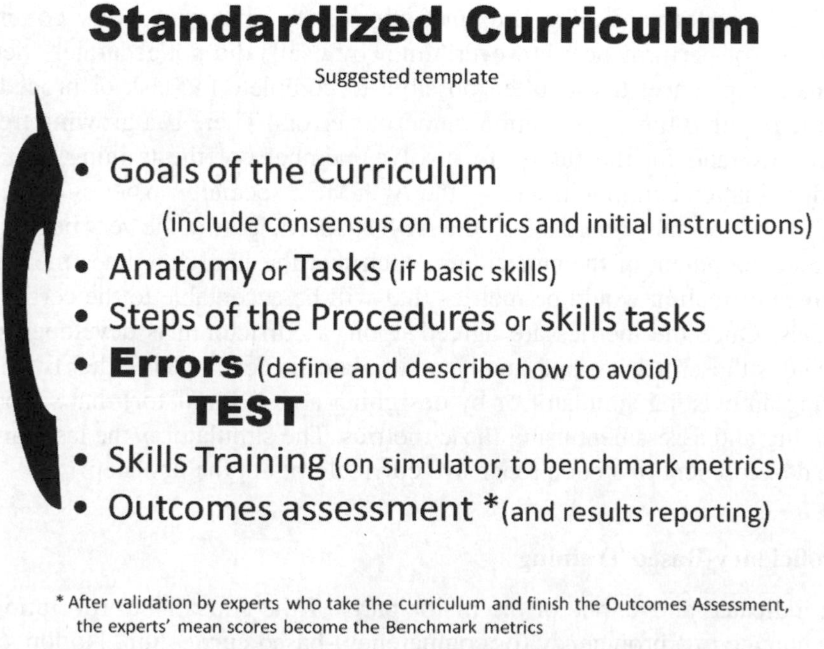

Standardized Curriculum

Suggested template

- **Goals of the Curriculum**
    (include consensus on metrics and initial instructions)
- **Anatomy** or **Tasks** (if basic skills)
- **Steps of the Procedures** or **skills tasks**
- **Errors** (define and describe how to avoid)
    ## TEST
- **Skills Training** (on simulator, to benchmark metrics)
- **Outcomes assessment** *(and results reporting)

* After validation by experts who take the curriculum and finish the Outcomes Assessment, the experts' mean scores become the Benchmark metrics

FIGURE 10.2   Curriculum template.

errors have been included in the initial didactic instruction: not only what the common errors for each procedure would be, but how to identify when one would occur, as well as techniques for avoiding making such errors. One additional requirement became evident—following the didactic portion of the curriculum, a test of this cognitive knowledge must be performed, and preferably with a score of 100%. The reasoning is that if the subject completes the didactic portion perfectly, if an error is made during the skills training, the error is a psychomotor skill error, since the subject has just demonstrated that he or she has full cognitive training. This permits the trainers to determine which skill component (cognitive vs. psychomotor) is responsible for the error, and to focus improvement efforts on either further knowledge acquisition or more psychomotor skills practice.

Objective Measures of Assessment

During the recent deliberations on mandating technical skills for board certification, the issue of correct outcomes metrics surfaced. Most curricula and simulators had been designed by an engineer from academia or industry, with the closest available "champion" surgeon to provide the clinical content. A significant number of the simulators developed did not have objective measures that were acceptable by the certifying authorities. In short, the outcomes measures of the curriculum did not properly assess the critical aspects of a subject's performance. For example, completion time was a very common measure of performance; however, time (by itself) did not accurately determine competence. It was often possible to complete the task or procedure in the required time, yet commit numerous errors. There is a growing trend, both now and for the future, to involve members of the training, testing, and evaluation communities (e.g., the ACGME), specialty societies (e.g., the ACS) and specialty boards (e.g., the ABS) to participate at the very beginning of a development of the curriculum, such that the final outcomes measures of the curriculum would be metrics that will be acceptable to the certifying bodies. Once the metrics are agreed upon, a curriculum is developed that includes those metrics, and finally, a simulator is developed (either by modifying an existing simulator or by designing a new simulator) that supports training and assessment using those metrics. The simulator is the last part of the development of a comprehensive curriculum, not the first step.

Proficiency-Based Training

As indicated above, one of the major parts of the educational revolution is the change to a proficiency (or competency)-based curriculum. No longer is it acceptable to train a subject on a specific skill for a fixed length of time (e.g., one day) and allow that subject to "pass" with an 80 to 85% score. This approach, in effect, is saying that the student or resident is permitted to begin treating patients with an expectation that approximately 15 to 20% of the time, he or she will make a mistake on a patient. Because there are numerous technologies (simulators) and quantitative processes (OSATS, OSCE, etc.) that can measure performance accurately, subjects must now perform to a benchmark criteria on: To be declared competent, they must score 100% (no errors) on two consecutive trials. When this level of proficiency has been obtained, the resident is allowed to operate on a patient, not before.

The question arises as to how the benchmark criterion is reached. The traditional method for determining a benchmark which is in practice today is to have expert or experienced surgeons perform the curriculum and simulation and be assessed. Their scores form the foundation for proficiency—the mean score ± one standard deviation becomes the benchmark for the task—and it is this benchmark that learners must achieve. Any score less than the benchmark is not acceptable, and a resident must continue to train until that benchmark is reached. Thus, the measurement of success is acquisition of proficiency (100%) to the level at which an experienced surgeon would perform, regardless of the number of trials or length of time. This is currently the best method to ensure the safety of the patient.

Validation Methodology

The acceptance of a curriculum (and simulator) depends not only on the proper outcomes measures and structured curriculum, but also the stringent validation of both the curriculum and the outcomes measures. Validation of the curriculum is easier to obtain because it can be carried out entirely under laboratory conditions (e.g., whether training reduces errors, time to completion, transfer from simulator to animal trial). Validation of the outcomes measures is more difficult because these metrics must be validated initially in the laboratory setting, but the ultimate outcome is whether the patient benefits (i.e., fewer complications, higher quality of life, etc.). With errors being infrequent to begin with, very large numbers of patients need to be studied to provide statistical validity, especially to determine whether proven improved performance (from laboratory training) will actually improve patient outcomes.

Standard statistical methodologies for validation have been available for decades, based on rigorous evaluation procedures developed by statisticians, behavioral psychologists, educational designers, and others. There needs to be a balance between rigorous science and practical application. Over the past decades, the validation measurements described below have become the accepted standards. ("Curriculum" here includes the outcomes measures, curriculum, and simulator.)

Face validity: ensures that the curriculum resembles the final task or procedure that it is meant to simulate—that there is a "face value" that is easily recognizable. This is performed during an initial consensus conference with subject matter experts and does not require quantitative metrics.

Content validity: ensures that the educational content (both didactic and psychomotor skills) contain the same tasks or skills that the real procedure would contain. Like face validity, this is part of the initial consensus conference.

Concurrent validity: ensures that the new, proposed curriculum is as effective or more effective in training than the currently available "gold standard" curriculum. This is measured by including the current standard as one arm of the validation design and comparing the results of each arm.

Construct validity: ensures that the curriculum is constructed such that it is possible to discriminate between a trained and an untrained person. This is usually carried out by having three groups: a novice (untrained) subject such as a medical student, an intermediate subject such as a resident, and an "expert" subject, such as a faculty member. The criterion for construct validity is met if the faculty performs better than the resident, and the resident performs better than the novice.

Predictive validity: ensures that the curriculum actually does improve performance, and that it is possible to predict which subjects will perform to criteria in clinical situations based on their performance during laboratory training. The caveat is that while it is validated that subjects who perform well in an operative procedure in the laboratory actually perform better during an operation on a patient (called *VR-OR training transfer*), there has been no long-term follow-up that shows that better performance in the operating room (OR) actually improves overall outcomes. As indicated above, long-term outcomes proof remains elusive.

A number of other measures, such as usability, reliability, and cost-effectiveness, also play into the ultimate successful commercialization (or requirement for certification) of a curriculum; however, these measures are beyond the actual scientific proof of the validity of a curriculum.

Transitioning from a Validated Curriculum to a Mandate for Certification

Just because a curriculum has been validated and published in a peer-reviewed journal does not mean that it will be adopted. There are numerous "targets" required for the curriculum to be adopted. The ultimate target is the

beneficiary, the patient who benefits from the surgeon's improved skills and reduction in errors. Another critical target is the **consumer**, the faculty who takes the curriculum and trains and assesses the learner. Perhaps the most common focus of the curriculum is the *user*, the student who actually uses the curriculum to train and upon whom the assessment is performed. And although the ultimate goal is patient safety, the primary goal of the user is to complete training satisfactorily so that they can become "board certified" to perform surgery or interventional procedures. Finally, there must be consideration of the *certifier*, the board that has the responsibility to ensure that the training truly does ensure the competency (and hence safety) of the surgeon. These must all be kept in mind as the curriculum moves from validation and publication to a product that is effective for all the levels involved. Figure 10.1 is used to exemplify the life-cycle management of curriculum: from initial consensus conference for outcomes to the final acceptance by the boards as a mandate for certification. Note that the governing authorities (the various boards and criteria-setting organizations) play the key roles from the onset of curriculum development until the final acceptance for certification. The example uses the pathway established for the SAGES/FLS course, which was adopted and mandated by the ABS.

If we accept as prima facie evidence that a curriculum is able to be designed to improve patient safety (otherwise, the curriculum is meaningless), each of the targets has described above specific needs, frequently different from the others. The faculty needs a curriculum that unambiguous and easy to use, simple for the student to understand, and has very clear and precise instructions, the result being a training experience that is easy to administer and unambiguous in the outcomes evaluation. There needs to be a very clear summative evaluation. The student needs a curriculum that is comprehensive for the task at hand, that explains in clear terms what needs to be performed (how), and the intended outcomes measures (what). There needs to be incentive, an ability to see progress toward succeeding at competence, with real-time feedback (formative assessment) of performance and the reward of successful completion (summative assessment). The certifier needs to be assured that the curriculum actually measures what it claims to measure, and that those measurements do reflect competency in the task to the point that it improves patient safety.

When all of the above are satisfied, the validated curriculum needs to be implemented. However, there are a number of steps that must be taken to ensure that the assessment is conducted properly. The ACS has instituted a certification program for simulation centers called **accredited education institutes** (AEIs). In an application and certification process similar to the one used by the RRC, the ACS-AEI surveys and certifies the quality of the simulation and training centers. This includes not only that there are adequate resources and personnel, but that the quality of training is acceptable, that not only are the students trained, but that the faculty are trained (training the trainers) and that outcome measures are continuously monitored and used to improve the entire spectrum of education. There are over 50 certified centers, which meet semiannually to develop curricula and learning management systems, and to conduct multiinstitutional research studies.

Once there is assurance that the curriculum is being used properly for training purposes, a system needs to be in place for testing for high-stakes assessment (i.e., acceptable to certifying authorities). One recent model has been the SAGES/FLS, which follows the model of the **National Board of Medical Examiners** (NBME). The SAGES/FLS receives applications from simulation centers to administer the FLS test. A surveyor is sent to assess the capabilities of the center and train the trainers to administer the test. This ensures proper conduct of the test, as well as the elimination of bias. The test results are then forwarded to SAGES/FLS for final grading and issuance of the certificate of successful completion of the FLS course. This is the certificate that is required by the ABS in the board certification process.

To ensure continuous improvement of the entire curriculum process, there is feedback regarding the metrics for the curriculum, based on factors such as testing results, changing the knowledge base, or new technology. Although curriculum development and building simulators are important, there must be attention to all the life-cycle details of the validation and certification process to make certain that future curricula meet the goals.

FUTURE TECHNOLOGIES AND APPLICATIONS

Although much about the revolution in medical education is being driven by modeling and simulation technologies, it is clear that simulation is a well-established discipline in most of the scientific, aviation, and military communities. Therefore, there will be significant familiarity with the discussion below. In fact, there is a direct analogy between existing nonmedical applications and the emerging opportunities for health care. The emphasis on team training is a direct adaptation of crew resource management from aviation. There will not be a review of the currently standard accepted simulations unless there are needs that could be fulfilled by a future discovery or adaptation.

The standard classification of simulation technologies is three categories:

1. *Live*, which includes actors (with or without moulage), manikins, simulated tissues, and animals (parenthetically, new laws mandating replacement of live tissue training by simulation will soon see the elimination of all animals in training).

2. *Virtual*, which includes VR and computer-based technologies: full VR immersive environments such as Second Life, VR task and procedural trainers, interactive Web-based training, and serious videogames.

3. *Constructive*, which includes planning tools such as mission (surgical) planning and rehearsal, networked simulators, and process models.

Although such a classification is arbitrary (as any classifications would be), this has been accepted for many decades by the simulation community at large and provides a "common language" to use in communicating characteristics of a simulation. Thus, there would be great advantage for interested stakeholders who are engaging in health care simulation to continue to use the classification for all future curriculum development.

Team Training and Continuity of Care

Pioneering work by David Gaba with his colleagues in aviation led to the first medical simulator, a manikin-based device initially used to train anesthesia skills, which was immediately adapted to intraoperative crisis management and has now evolved into the standard method of team training [2]. Recent advances have included interdisciplinary team training, and in situ training [e.g., in the emergency room (ER) or intensive care unit (ICU)] and even testing of team skills. The level of physiological sophistication continues to increase, and although the current manikins look much more lifelike than their predecessors did, there is a great need for a manikin that not only appears nearly realistic but can also manifest many more of the critical physiological factors, such as bleeding and tissue responsivity. A number of simple procedures can be performed on a manikin, such as chest tube insertion or needle tracheostomy, although the number of procedures and fidelity of the tissues need to be increased by orders of magnitude. Performance analysis (postprocedure debriefing) is very tedious and time consuming, and suffers from a high degree of subjectivity and poor interrater reliability. Some of this could be improved with more stringent curricula, but automatic image analysis, salient feature recognition, and automatic tracking could greatly improve the current status. The holy grail would be to have an automatic analysis system that could recognize errors and provide immediate feedback—essence, a virtual mentor.

Team training has made remarkable strides forward in developing communication skills, but the actual "choreography" of an emergency room trauma evaluation, OR crisis, or ICU resuscitation has not been addressed. There are no "standard" positions around the patient for the participants to take, and the physical changing of positions, passing of instruments or equipment, and similar details have yet to be attempted. Communication and professional skills are well instructed and assessed, but the actual mechanics of interacting physically with other members of the team need to be choreographed. All other professions carefully choreograph members of the team, whether in sports (playbooks with diagrams), dance, theater, or the military.

The next step beyond team training is continuity of care: where one team "hands off" the patient to the next team. This has been identified as a common source of error in patient care [9]. Too often the correct equipment or connector is not available at the destination, or the receiving team does not get the critical information about care given previously, laboratory or imaging data, or medications administered. A few scenarios have been demonstrated, such as from ER to OR to postoperative recovery, or ER to delivery room to neonatal ICU, or correct transport and transfer to an imaging facility. In the military, this has been referred to as the *chain of evacuation* [10]. Numerous scenarios are needed to reproduce the difficult prehospital environment, variety of evacuation systems (ambulances and aircraft), and types of equipment and exchanges at every level of care.

Clinical Examination: Virtual Patients and Virtual Cadavers

The clinical exam, which consists of the history and physical exam, demands a large amount of knowledge, communication, and professional skills. The current method is to train professional actors to be patient actors, representing a significant commitment in time, money, and resources. There are unpublished reports of the development of **virtual patients** (actors). The limitation with today's real actors is that they can provide only a limited number of (scripted) responses when mimicking a disease state. However, a virtual patient can be programmed to have the full spectrum of a disease, such as diabetes, and a student could continue to "examine" the virtual patient time and again until all the various manifestations of the disease have been presented. Then when examining the real actor, the student will have a much broad and deeper understanding of the disease process. Research is needed in the field of **human social, behavioral, and cultural** (HSBC) characteristics so that the virtual patient could demonstrate the appropriate emotional, facial expression, pose, and gestures to represent not only their current emotional state, but also cultural effects that greatly influence a physician's interaction with a patient. The visual fidelity of virtual people in the entertainment industry is so good that it is nearly impossible to distinguish a real from a virtual actor; however, the limitation is that these virtual actors do not have physiological or biological embedded properties, nor do they respond spontaneously to questions or physical examination. Even though there is a great distance to go before virtual actors both look and behave realistically, enough basic science exists to provide a reasonable facsimile that would enable practice by a novice medical student or nurse.

One of the most important concepts is that virtual patients provide the opportunity to develop a **digital library of patients**. This is an enormous opportunity and benefit for students. The apprenticeship model used today relies on "the patient who comes through the door," in order for the student to gain clinical experience. If a patient with a critical common disease does not happen to be admitted to the service during the student's rotation, it will not be possible for that student to gain the necessary experience. This is becoming an even more acute problem with the initiation of the 80-hour workweek (and the European 56-hour work time directive), with the likelihood of even more severe restrictions on a student's or resident's exposure to patients. The good news is for simulation companies, since this is one business model that provides an ongoing revenue stream: which can create dozens or even hundreds of variations of disease states, as well as virtual actors to portray these diseases. As technology improves, there will be an even greater demand for more and more sophisticated virtual patients.

The equivalent to the virtual patient for the first-year medical student will be the **virtual cadaver**. Just as students experience only patients who are admitted to a specific clinical service, the medical student has the opportunity to learn fundamental anatomy only on the cadaver (shared by four students) to which they are assigned. Once a dissection is performed, the other three students are not able to dissect the cadaver, and frequently critical structures are destroyed in the course of the dissection. When studying the genitourinary tract or breast disease, the students must "double up" with students who have a cadaver of the appropriate gender. Similarly, there is very little opportunity to have exposure to the anatomy and structure of the major disease states, since very few cadavers actually have a major disease (or had organs removed in an attempt to cure the disease). The University of Washington is beginning a "virtual cadaver and dissection" program based on CT scans of a student's own cadaver. The virtual cadaver can be dissected in an unlimited number of ways and times. Eventually, the anatomy program will be able to collect a "virtual library of cadavers with diseases or abnormalities," although industry may want to sell virtual cadavers to supplement the anatomy program's library. The advantage is that once there are a sufficient number of cadavers (or for clinical experience, virtual patients), it will be possible to develop a comprehensive structured course of instruction which includes all of the important (and even little known) diseases with which the student must be familiar, and provide the opportunity not only to dissect (or practice surgery on) the disease state, but also the opportunity to conduct a clinical exam on the same (matched) patient.

In the extremely long future, each person will have a total body scan of their own body and will embed all their data (biological, vital signs, genetic, physiological, etc.) in the image, which then will become a computer surrogate (a "body double") for them. This concept, referred to as a *holographic medical electronic representation*, was developed originally by the Defense Advanced Research Projects Agency as the Virtual Soldier Program [11]. Ultimately, sophisticated "predictive algorithms" will be developed to allow a person to determine what the consequences of a certain medication, diet, or behavior would be on their long-term health. This may become a realization of "The Picture of Dorian Gray," aging your image to determine the results. Since understanding of the human body is making remarkable strides every day, the building of holomers and predictive (simulation) tools will be a continuing endeavor and business opportunity.

Basic Skills and Full-Procedure Simulations

When developing curricula for skills, whether individual tasks or full procedures, the most important issue is to match the fidelity of the simulation to the level of the learner and complexity of the task or procedure. For example, a photo-realistic simulator is not needed for a very basic skill such as a fundamental pick-and-place exercise (e.g., rings on a peg) for a beginner or a peripheral intravenous catheter insertion. There are excellent basic skills curricula that have been well validated and even required for certification, such as the SAGES/FLS curriculum; what are needed are a number of the simple procedures that are performed using an "open" technique (as opposed to a video-based technique). These are needed for prehospital care (e.g., hemorrhage control, chest tube insertion, open tracheostomy) and other skills from both the surgeons' "Advanced Trauma and Life Support" and the medics' "Pre-hospital Trauma and Life Support" manuals. A few such simulators have been created; however, numerous compromises have been made in order to create a simulation that supports the curriculum. Whereas the tasks cited above are rather straightforward and simple, recreating a high-fidelity manikin or simulated tissue model for such skills has been very challenging and so far has eluded the best of engineers. One of the key challenges is to develop simulated tissue, based on biological principles (most simulated tissues are based on physics engineering principles), which "behaves" as real tissue, including bleeding, twitching, separation of tissue planes, and so on.

In addition to manikin and simulated tissue, there are a few hybrid simulators which combine VR images on a monitor screen for the visual effect and for real-time illustration, and a simulated tissue model for the haptic component of real tissue. Conducting full, complicated intraabdominal procedures on a manikin or simulated tissue torso has proven difficult because of the complexity of the numerous structures involved. At this time, even if such a simulator were created, it is highly likely that the cost would be prohibitive. Thus, the future for manikins and simulated tissue as well as for hybrid simulators remains very challenging, with a full spectrum of opportunities, from simple tasks to full procedures that can be produced at an affordable price.

Cost is a very important aspect, since tasks and procedure training on a live object destroys the product, limiting the opportunity to conduct the training to one or a few trials, after which replacements are needed. Since the goal of simulated tissue is to mimic real human tissues and organs, will it be possible in the future to actually develop "self-healing" surrogate tissues? Or will the day come when synthetic organs are specifically grown (as is currently being done with tissue engineering) inexpensively enough for training? Since the complexities of transplanting organs into a live patient will not be required, it may be possible to make such synthetic living tissue cost-effective.

Given the direction of regulations and loss of animal training models, described above, there will be an increasing need for VR and computer-based curricula. Future systems will need to focus not only on the standardization of content (which is acceptable to certifying authorities) but also on a major increase in the fidelity of all types of simulations. Physical models (synthetic tissues, manikins, etc.) as well as virtual models will need not only to look real, but also to "feel" real and be able to be dissected in a natural fashion. There must be bleeding, muscle twitching, correct plasticity of bones, and so on, based on measured biological properties. Taking simulation from simple skills and tasks to full operations will require extremely complex systems of systems engineering to integrate not only the local interaction between tissues and organs, but also the systemic responses related to neurologic, hormonal, and other signaling factors. Then, once an initial model is created, it will need to be parsed into literally thousands of variations, based on diseases, congenital anomalies, common normal variations, and so on, in order to provide a very wide range of alternatives to practice on. The ideal methodology would be to begin with variations discovered on CT scans, so that the individual anatomy of patients can be imported into the simulator (see the discussion below of pre-op planning and surgical rehearsal). Virtual models have advantages as well as disadvantages. The greatest advantage is that the virtual model simulation can import the image from CT or MRI scans, and create an entire digital library of innumerable variety quickly and inexpensively (perhaps CT scans can be exported to stereolithography machines to create tissue surrogates, or to cellular bioprinters to print synthetically grown organs). Sophisticated "authoring programs" can permit local educators to quickly change parameters, both structural and functional (physiology, biochemistry, etc.) in order to build a curriculum that suits local needs while still meeting national requirements. On the other hand, the science of haptics remains in its infancy, and new discoveries will be needed to bring virtual haptics (for a

VR environment) to a level that is even close to reality. Thus, the future of simulators per se is very bright, with very large opportunities.

Clinical Applications

Nearly all the curricula that have been developed to date have focused on training and assessment in the safe environment of a simulation center. However, there is the beginning of migrating the training and assessment into the clinical setting, as either a staged training scenario or as a routine part of patient care.

In situ Training   This is a method of taking the actual simulator (usually, a manikin) into the hospital and initiating a "crisis" (usually a "code" for cardiac arrest) in the actual patient bed on the ward or specialty unit (such as an ICU or ER) without a respondent realizing that it is a training (or assessment) opportunity. The team must then "save" the patient; for those situations where the crisis management scenario is for assessment, if the team does not save the patient, they will need to undergo remediation retraining in the simulation center before being eligible to participate in the crisis management of real patients.

Preoperative Warm-up   Initial studies by Kahol et al. [12] have demonstrated unequivocally the advantage of a 15-minute warm-up exercise on a simple task simulator which addresses all the critical cognitive and psychomotor skills. The concept is to activate the attention center in the brain to increase concentration, short-term memory, and working memory as well as to stimulate the critical psychomotor skills of precision, hand–eye coordination, visiospatial orientation, and depth perception. Although the tasks are quite abstract and simple, the design is so efficient and focused that all the critical areas are addressed. Future curricula will need to be more sophisticated as further understanding of critical training skills is unearthed. Research into the duration of the effect of warm-up as well as critical factors affecting the warm-up, such as noise distraction and fatigue, is essential. Finally, decisions are needed regarding requiring pre-op warm-up before every procedure, placement of the simulator (e.g., in the OR?), and so on.

Preoperative Planning and Surgical Rehearsal   It is inconceivable that a pilot would begin a flight or that a military unit would conduct a combat mission without careful and rigorous planning and rehearsal, yet surgeons will dash into an operating theater without rigorous planning of a procedure, let alone rehearsing such a procedure. However, simulation technologies provide the same opportunities as in other professions to very meticulously plan and rehearse a surgical procedure on a patient-specific CT scan in a simulator. The first successful simulation systems to include patient-specific images for practice before surgery have been the endovascular simulators. Mutter et al. have been rehearsing complicated liver surgery since 2008 [13]. Considering the rapid pace of technology, it is very likely that in the near future, surgical rehearsal will be an essential (if not a required) part of every complicated surgical procedure. Interestingly, it is also scientifically possible to use an abbreviated form of surgical rehearsal when a surgeon describes a required surgical procedure to a patient as part of the consent for surgery.

Cyber therapy Virtual environments have been used in a number of areas, specifically in anesthesia (for pain management) and in psychiatry. Distracting a patient (especially children) by an engaging virtual world or game when undergoing a surgical procedure or wound debridement can lower the amount of pain medication required. Immersing a patient who has a phobia in a simulation of the stimulus causing the phobia, such as height, open spaces, or flying, and desensitizing the patient without exposing the person to risks, has been successful in many areas. One application critical to the military is the use of VR to diagnose and treat traumatic brain injury and posttraumatic stress disorder [14]. This is expanding beyond the hospital and clinic and becoming available through Second Life (and other virtual worlds) to provide access to treatment or even just camaraderie, not only for returning soldiers, but also for support groups for cancer and other diseases.

Administrative Integration Whether in the simulation center, in situ environment, or as a part of clinical practice, simulation data are either kept in the simulation center database or are not even collected. However, results of the training and assessment, of pre-op warm-up, planning, and rehearsal, and of patient instruction have enormous potential to improve overall patient safety through the administrative process. Results of team training and continuity of care could be shared with the risk management and hospital efficiency committees, including improvement in OR teams and room turnover times. Initial training and assessment scores can be shared with the hospital privileges committee as objective measures of performance to ensure the competency of new physicians. The same information can be acquired during reassessment for maintenance of certification and retraining for a new surgical procedure, all part of the re-credentialing process. Analysis of the performance of those taking the training could be forwarded to the risk management committee for that committee to be aware of the procedures and physicians most likely to cause errors and hence require further training. The information from the database could also be shared with the patient safety committee. In the future, a surgeon would download a patient's specific three-dimensional image (CT, MRI), and illustrate on the image exactly how a surgical procedure would be performed. Such a session could be saved digitally and become part of the consent form. If the surgeon conducts a preoperative plan and surgical rehearsal, the data can be sent to the central supply department, and the information about the instruments and supplies needed for that simulated case could be used to ensure that the surgeon has exactly the instruments and supplies that are required for surgery. These are just a few examples of how sharing information from training and assessment sessions could benefit the management of patient care and decrease risk while improving hospital efficiency. To meet these and the numerous other information technology demands, the hospital of the future may need its own supercomputer and limitless data storage capacity.

CONCLUSION

Modeling and simulation are just beginning an upsurge in the field of health care. The fundamental issue is the development of the curriculum in the comprehensive sense of the term. This means that modeling and simulation must be a supporting part of the entire curriculum, which begins with a consensus conference that determines the final outcomes measures that are acceptable to certifying authorities. Then the didactic portion of the curriculum is developed and must include unambiguous descriptions of common errors and how to avoid them. Once the pedagogical portion is complete, the appropriate simulation can be chosen, whether adapted from existing simulators or created specifically to support the new curriculum. The simulation needs to support not only the cognitive and psychomotor aspects of training, but all the assessment tools (both formative and summative), which allows for automated, objective results reporting. For simulations of patients, all the foregoing aspects, plus the HSBC and emotional components, must be included. These processes are critical to the future of modeling and simulation development.

Also, numerous opportunities are provided by new technologies and applications, many of which are already standard in other disciplines and can be adapted rapidly to health care. While there are a number of near-term opportunities involving virtual patients, digital libraries, serious videogames, in situ training, preoperative warm-up, preoperative planning, and surgical rehearsal, there remain some serious challenges to increasing the visual fidelity, creating simulations that are biology-based rather than physics-based, developing full manikin simulators with biorealistic tissues, or perhaps even growing specific organs and tissues to replace the use of animals. Finally, modeling and simulation must not only be limited to a laboratory setting, but must move out into the clinical environment and become a common tool for direct patient care and supportive of hospital administration. Bringing together expertise from government, academia, medical governing bodies, and industry will provide a strong and viable platform for the future of medical modeling and simulation.

KEY TERMS

Residency Review
Committee
American Council of
Graduate Medical
Education
American College of
Surgeons
Society of American
Gastrointestinal and
Endoscopic Surgery
objective structured
clinical examination

objective structured
assessment of
technical skills
face validity
content validity
concurrent validity
construct validity
predictive validity
beneficiary
consumer
accredited education
institute
National Board of
Medical Examiners

virtual patient
human social
behavioral, and cultural
characteristics
digital library of
patients
virtual cadaver
in situ training
preoperative warm-up
preoperative planning
and surgical
rehearsal
cyber therapy
administrative
integration

REFERENCES

[1] Flexner A. Medical Education in the United States and Canada: A Report to the Carnegie Foundation for the Advancement of Teaching. 1910. Original document: http://www.carnegiefoundation.org/files/elibrary/flexner_report.pdf. Accessed Nov. 19, 2009.

[2] Gaba DM, DeAnda A. A comprehensive anesthesia simulation environment: re-creating the operating room for research and training. *Anesthesiology*, 69(3):387–394, Sept. 1988.

[3] Satava RM. Virtual reality surgical simulator: the first steps. *Surg Endosc*, 7:203–205, 1993.

[4] Resnick R, Regehr G, MacRae H, Martin J, McCulloch W. Testing technical skill via an innovative bench station examination. *Am J Surg*, 173:226–230, 1997.

[5] Derossis AM, Fried GM, Abrahamowicz M, Sigman HH, Barkun JS, Meakins JL. Development of a model of evaluation and training of laparoscopic skills. *Am J Surg*, 175:482–487, 1998.

[6] Seymour NE, Gallagher AG, Roman SA, O'Brien MK, Bansal VK, Andersen D, Satava RM. Virtual reality training improves operating room performance: results of a randomized, double-blinded study. *Ann Surg*, 236:458–464, 2002.

[7] Seymour NE, Gallagher AG, Roman SA, O'Brien MK, Bansal VK, Andersen D, Satava RM. American Council on Graduate Medical Education (ACGME). Program Requirements for Graduate Medical Education in Surgery: Common Program Requirement, Effective: January 1, 2008, Section II D (2), 2009. http://www.acgme.org/acWebsite/downloads/RRC_progReq/440_general_surgery_01012008_u08102008.pdf, p. 10. Accessed Oct. 19, 2009.

[8] Fundamentals of Laparoscopic Surgery. http://www.facs.org/education/fundamentalsofsurgery.html. Accessed Oct. 19, 2009.

[9] Seymour NE, Gallagher AG, Roman SA, O'Brien MK, Bansal VK, Andersen D, Satava RM. *To Err Is Human: Building a Safer Health System*. Washington DC: National Academy Press, Nov. 1999.

[10] Turner S, Ruth M, Tipping R. Critical Care Air Support Teams and Deployed Intensive Care. *J R Army Med Corps*, 155(2):171–174, June 2009.

[11] Satava, R. Innovative technologies: the information age and the bioIntelligence age. *Surg Endosc Ultrasound Intervent Tech*, 14:417–418, 2000.

[12] Kahol K, Satava RM, Ferrara J, Smith ML. Effect of short-term pretrial practice on surgical proficiency in simulated environments: a randomized trial of the "preoperative warm-up" effect. *J Am Coll Surg*, 208(2):255–268, 2009.

[13] Mutter D, Dallemagne B, Bailey C, Soler L, Marescaux J. 3-D virtual reality and selective vascular control for laparoscopic left hepatic lobectomy. *Surg Endosc*, 23:432–435, 2009.

[14] Rizzo AA, Difede J, Rothbaum BO, Johnston S, McLay RN, Reger G, Gahm G, Parsons T, Graap K, Pair J. VR PTSD exposure therapy results with active duty OIF/OEF combatants. *Stud Health Technol Inf*, 142:277–282, 2009.

APPENDIX
Modeling Human Behavior, Modeling Human Systems: Addressing the Skepticism, Responding to the Reservations

JOHN A. SOKOLOWSKI and CATHERINE M. BANKS

Abstract

Advances in modeling and simulation (M&S) allow for simulation of complex phenomena such as human behavior modeling and the modeling of human systems. Human Behavioral Modeling allows for the incorporation of socially dependent aspects of behavior that occur when multiple individuals are together. Human Systems Modeling now includes more complete models of human physiology with higher fidelity, more realistic simulations. Juxtaposed to these technical advances is skepticism about representing human behavior and reservations about simulating physiology. These disclaimers are stemming from disciplines that rely heavily on soft (or fuzzy) and evolving data—the social sciences and medical/health care fields of study. Both are concerns that the developers (modelers and simulationists) must address to promote modeling and simulation as a multi-disciplinary tool. The purpose of this paper is to address the unease of using modeling and simulation in the representation of human behavior and human systems.

INTRODUCTION

Advancements in computer software and hardware and artificial intelligence and software agents have hastened the pace of modeling and have enhanced

Modeling and Simulation in the Medical and Health Sciences, First Edition. Edited by John A. Sokolowski and Catherine M. Banks.

195

the capabilities of simulation for characterizing and representing more complex phenomena such as the human personality in social and conflictual simulations – modeling human behavior; and complex physiology – modeling human systems.

A behavioral model is a model of human activity in which individual or group behaviors are derived from the psychological or social aspects of humans. Behavioral modeling allows for the incorporation of socially dependent aspects of behavior that occur when multiple individuals are together. This type of modeling is now being used in fields of study that include observations of human behavior be they individual, group, or crowd behaviors (Banks, History 2009).

The modeling of human systems is also very mature. From the introduction of modeling human systems using mannequins (1960s) to the more complete models of human physiology, higher fidelity and more realistic simulations are now being developed. Medical simulation is becoming an accepted methodology for educating future medical practitioners and for providing ongoing training and assessment for practicing professionals. Effective use of simulation promises to help medical and health care personnel become more productive and to leverage their teaching time across a larger number of students (Combs, 2009).

Thus, modeling and simulation (M&S) is used to represent human behavior and human systems as these are fields where experimentation is conducted using dynamic models. M&S is distinct in that it is often the only tool capable of analyzing complex systems because it allows for an understanding of system dynamics and it includes enabling technology both of which provides a means to explore credible analysis. Furthermore, there is an increasing awareness of the need to integrate soft data into models characterizing real-world phenomena and actual human anatomies. Granted, representing human society and individual relationships in that society, as well as individual relationships to that society is technically challenging, it can be done (for examples see Sokolowski and Banks, Real-World 2009). However, these and other characterizations of human behavior give rise to skepticism among the social sciences. Among the medical and health sciences there are reservations as to the accuracy of modeling human systems as the technology strives to represent the human body and its vast array of minute variations.

The purpose of this paper is to address the skepticism and reservations relative to using simulation in the representation of human behavior and human systems. These disclaimers are stemming from disciplines that rely heavily on soft (or fuzzy) and evolving data—the social sciences and medical/health care fields of study. Developers (modelers and simulationist) must respond to this unease of using simulation to promote M&S as a multidisciplinary tool.

At the core of the discussion are the questions: Is simulation a valid and verifiable means of representing humans? How can human actions (behavior) and human physiology (systems) be reflected in computer models? And, how is that accuracy ascertained? Interestingly, the nature of these questions is not new to M&S professionals as they have been addressed in a different context. Thus, a good starting point for this discussion is in the community of the developers, the M&S professionals, and the standards by which they operate.

A CODE OF ETHICS WITHIN THE M&S COMMUNITY

Recall the fundamental difference between modeling and simulation: modeling deals primarily with the relationships between real-systems and their representations, while simulation refers primarily to the relationships between computers and models. To conduct prescriptive or predictive analyses, one cannot be had without the other. And it is the prescriptive/predictive potential that incites skepticism and invites reservation.

In fact, there is a sizable body of literature on the question of ethics in M&S and it spans various disciplines: business ethics, computer and information ethics, defense ethics, science ethics, and engineering ethics. The first dialogue was had by the engineers themselves when in 1983 John McLeod, editor of Simulation published by Sage, began formal discussions on the subject. By 1985 a Code of Ethics was being contemplated.[1] McLeod's concern was from a developer's perspective—the credibility or validity of the model—his intent was to ensure no manipulation of the model.

At that time the concern for the M&S community centered how the results of any assessment or model can be drastically changed or even reversed by small changes in the values of soft data or by assumptions concerning details of the structure of the model. To avert an absence of ethics McLeod included in the *Guidelines* responsibilities to various members of the M&S community: the responsibility to oneself, to one's colleagues, to one's client, to society, and to one's profession. Specifically stated in the section Responsibilities *to One's Profession* are the following admonitions: reliance on principles of good computer M&S practice remembering component relationships and assumptions that determine the form of the model while numerical data determine the results beware of *garbage in/gospel out attitude* on the part of client who is more concerned with results; moreover, avoid the use of *soft data* if at all possible (McLeod, But Mr. President 1986).

---

[1]Experts who participated included: WJ Garland, D Radell, JG Stevens, L Kramer, RG Sargeant, S Schlesinger, CA Pratt. The M&S professionals who developed the finalized Code included : TI Oren, S Elzas, I Smit, and G Birta.

For two decades there has been an ongoing dialogue on the ethics of simulation from developer community. These discussions have culminated in a formalized *Code of Professional Ethics for Simulationists* developed by a body of M&S professionals in 2002 (Oren, Code 2002). The Code includes verbiage regarding personal development and the profession, professional competence, and trustworthiness. Like all codes of conduct, it is up to the individual to adhere to its precepts. What is significant, however, is that this body of professionals has put into place a means to maintain a high level of ethical conduct. Additionally, the code is not the only measure to ensure credible model development. The discipline of M&S has an inherent system of checks and balances – verification and validation (V&V).[2]

There are many case studies which lack adequate social theories to characterize social systems. The same goes for human behavior. With no underlying explanation about patterns of behavior the M&S professional must rely on anecdotal or empirical evidence on which to base his model. The model developer has an ethical responsibility to recognize this limitation and to clearly state on what assumptions his model is based. This process at least alerts the users of the model's potential inaccuracy.

SKEPTICISM IN THE SOCIAL SCIENCES

The question of simulation use among social scientists is as broad as there are individual disciplines. Those in the business studies often think in literal terms of ethics in that workers are not equipped with the best tools for fully evaluating ethical decisions and they rely on gut reaction or on personal values. Along those same lines is a serious discussion in the disciplines of education, psychology, and communications on the relationship between user behavior and digital representations (simulations) in interactive entertainment. This is premised on the precept social behaviors are often learned without conscious intellectual understanding concluding that the way one rationalizes or explains an activity intellectually can be diametrically opposed to the behaviors that have been learned and enacted. As social science modelers seek to develop real-world depictions in their simulations they realize the real-world representations require characterizing human behavior. To develop such models qualitative data (aka soft, squishy, or fuzzy) is necessary. A goodly number of social scientists are troubled with the use of this type of

---

[2]The V&V process is critical to ensuring that a simulation is an accurate representation of its real-world counterpart. Failure to carry out this process jeopardizes simulation credibility and leaves to chance its accuracy. M&S practitioners have developed various V&V methodologies that have formalized this process (Balci, 1998 and Petty, 2010).

data. They question how to translate or map data, and wonder if the "mapped" data is simply a distillation of human behavior to a series of computations. This causes doubt as to the certainty that these models accurately reflect (or reasonably approximate) human behavior. Moreover, how can the accuracy of a simulation depicting human behavior be ascertained? Therein lies the skepticism.

Keep in mind that representing humans is in essence the modeling of many systems or a complex system packed with soft data.[3] Realistically addressing the issue of modeling soft data, McLeod notes that the inclusion of soft data, especially when used in case studies that serves as adjuncts to decision-making, imposes an increasing responsibility on the part of modelers and analysts. As such the question of data selection, data input, judgment of modeler, judgment of analyst, and the modeler's interpretation of dynamics of the system will affect simulation output. Often these projects or case studies must determine and disclose to decision-makers the extent to which information derived from the study is valid and applicable to the decision of interest (McLeod, Professional 1983). And this type of modeling is new to many M&S professionals who have for some time developed models of physical systems whereby the characteristics of the objects being modeled are known or can be ascertained, no soft data involved. There is an increasing awareness among modelers of the need for soft data representation. If the method of evaluating the results of a study can change the conclusions, then it stands to reason the choices involved in M&S, who makes them, and under what influences, raises an important ethical question. Still, the question of ethics from the developer, M&S professional, standpoint is one that has been settled and codified. The *M&S Code of Ethics* is the reassurance that M&S has an inherent and prescribed conduct in which the user community can reference and rely on.

RESERVATIONS IN THE MEDICAL AND HEALTH SCIENCES

The discipline of M&S and the use of M&S applications is grounded primarily on analysis, experimentation, and training. Analysis refers to the investigation of the model's behavior. Experimentation occurs when the behavior of the model changes under conditions that exceed the design boundaries of the model. Training is the development of knowledge, skills, and abilities obtained as one operates the system represented by the model. With this multi-dimensional capability M&S is credited with being an enabling

---

[3] Soft data is defined as information bearing on a decision that is difficult to quantify, it might be obtained from questionable sources, it can change with time in unpredictable ways, and it could be inadvertently skewed by the paradigm of the collector or user.

technology which has been making large inroads among the medical and health sciences (MHS). Various simulation societies focused on medical applications speak to the emerging trend of engaging simulation in differing modes such as live, virtual, and constructive simulation. The subfields of MHS make use of simulation from training with mannequins to using haptic devices to imaging devices and virtual operating rooms.[4]

There is a plethora of applications with promise of even more as medical simulation is used in multiple corners: research, evidence-based outcomes, medical education, performance assessment to name a few.

In general there are three types of medical simulation, or surrogacy as some prefer to call it, in the clinical setting: human, mechanical, and virtual (Bauer, 2006). *Human simulation* uses a trained role-player to act the part of a patient with a specific medical condition. (The role-players are often professional actors.) This has been problematic in that a major limitation of human simulation is the inability for students to perform invasive procedures and other therapeutic interventions that could be harmful to the role-player. *Mechanical simulation* allows students to use mock or artificial parts to adequately mimic the experience that would be gained from interacting with a real patient's body, organs, or tissues. *Virtual simulation* employs the latest advances in computer technology and visual interfaces to create acceptably realistic learning experiences. Game-based medical simulation is making advances proving to be an efficient and effective tool for teaching many clinical lessons without real patients and instructors. Bound up in all three modes is an implied trade-off between reality and simulation, but does not necessarily mean that one is superior to the other (Bauer, 2006). So what role should simulation have in MHS?

In a 2001 *Simulation Gaming symposium* an assessment of state of affairs in medical M&S was discussed (Crookall and Zhou, 2001). Included was the general concurrence that the bridges between medicine and simulation are many and varied. Medical simulation is in overlapping areas of medicine and healthcare. And this for many practitioners is a good thing as the increase in world population, improved awareness of health issues, aging of society, and medical progress itself is causing a quantitatively and qualitatively increase in healthcare demands. Thus, the symposium concluded that there is a need to exploit the benefits of simulation as a tool to train and to engage (in procedure).

Medical M&S is grounded in the engineering and computer science disciplines where computational intelligence applications are being applied to cardiology, electromyography, electrocephalography, movement science, and

---

[4]Haptic technology serves as the interface within a simulated environment engaging the user's sense of touch by applying forces, vibrations, and/or motions to the user.

biomechanics. Bio-medical engineering has also made inroads by engaging engineering concepts and techniques to typical applications in prosthetics, medical instruments, diagnostic software, imaging equipment (Begg, 2008).

Pedagogically, medical M&S it is being introduced with great haste as a teaching tool. Along with the technology specific to health care is the need for doctors and healthcare providers to gain and maintain competence and demonstrate proficiency in the use of these new technologies. This makes a compelling case for the increased use of simulation technologies in medical education and medical practice. Still, this type of education and training is not without its reservations by those who raise questions of model and simulation validity.

For example, there are numerous potential applications of simulation for the assessment of clinical competence; however, it has not been widely supported within the healthcare community (Riley, 2008). Concerns about the application of simulation to performance assessment include competing tensions between dual goals: one goal is to achieve high reliability and the other goal is to achieve high validity. The concern is that these goals appear to require mutually exclusive test conditions. Still, experimental studies have demonstrated that moderate levels of reliability and validity can be achieved simultaneously with mannequin simulations if test conditions are appropriately managed. Some have made the case that simulation not be used in isolation to make a determination about a practitioners overall performance; rather, it should be incorporated into a broader multi-faceted program (Riley, 2008). For healthcare educators, simulations allow the practice of safety, prevention, containment, treatment, and procedure in a risk free setting (Kyle, 2008). A prominent supporter of simulation in healthcare education is Dr. Richard Satava of the University of Washington, Department of Surgery. Satava believes simulation is an appropriate medical educational tool. He legitimizes this statement by citing the Residency Review Committee of Accreditation Council on Graduate Medical Education which has begun requiring residency programs to have simulation as an integral part of their training programs. The American College of Surgeons (ACS) is also making use of the tool and it is certifying training centers to ensure the quality of training provided.[5] Simulation creates an evidence-based learning environment as it allows for anything that reproduces experimental, clinical, or educational data. The model is arguably the most important part of the simulation as it constrains the simulation from being a false representative.

Thus, clinical simulators need a model to drive them. Herein lies the reservation because a simulation is no better than the model it engages. In

[5] As of 2008 the American College of Surgeons certification requires three categories of students that are to be taught using simulation.

the MHS, model development has been questioned. Simply, if the physiology isn't realistic, the simulation isn't realistic. Sufficed to say M&S professionals realize that even the best model is a compromise as not every detail can be included and model compromise is an art and it must integrate input from subject matter experts.

MAKING IT REAL: ADDRESSING THE SKEPTICISM, RESPONDING TO THE RESERVATIONS

To address the concerns of M&S in the social sciences and medical and health sciences, the M&S community of developers must collaborate with the users as population plays a significant role in assessing human behavior or representing human systems. This at-large community is comprised of 1) the modelers, who select (or are given) data and make assumptions on the content/characterization of the model; 2) the simulationists, who design the experiment (or training scenario); and 3) the analysts, who interpret the results of the simulation. It is important to note that the modeler is dependent on data to develop his model and the best data in any model design comes from subject matter experts. Thus, models representing human behavior, such as decision-making or human interaction, rely on the qualitative analysis of subject matter experts who study human behavior such as social scientists.

For the most part, social scientists who are trained in modeling and simulation are done so from the social science approach. This includes the three common social science modeling techniques: *statistical modeling, game theory,* and *agent-based modeling.* These modeling paradigms are well-established and proven effective; however, they lack the ability to represent more complex systems. M&S professionals implement *systems-based approaches* as a way of developing a wholly inclusive characterization or representation from the macro-level (system dynamics) to the unit of analysis (agent) or node (game theory).

Representing the human systems requires expertise in the study of the anatomy such as those in the MHS. For professionals in the medical and health sciences, there appears to be less skepticism and more reservation with simulation use. This is no doubt partly due to the fact these professionals have no training in M&S as developers. Additionally, there is a hesitancy to use any tool that does not provide a mirror-image of the human physiology. This requires that simulationists recognize the parameters and capability of a model for the creation of an experiment or patient case-study. Thus, a modeler or simulationist skill-set should include subject matter expertise. Additionally, the analysts (users) who interpret the results of the simulation should also have subject matter expertise as well as and understanding of how the model was developed, its design, intent, and limitations. This facilitates their ability to review the simulation results to prescribe a response. The inclusion of medical subject matter expertise in the development of the model and design of the simulation will provide the accuracy that can moderate reservations.

Critical to model creation and simulation development is obtaining dependable results. This includes the development of techniques to firm-up soft data and the representation of that soft data. *In Models, Measurement, and Computer Simulation: the Changing Face of Experimentation*, Margaret Morrison argues that computer simulations have the same epistemic status as experimental measurement when looking at the role models play in experimental activity, particularly measurement. Morrison defines models as measuring instruments and simulations as the experimental activities. A model can be based in some theoretical belief about the world that is suggested by the data, or it can sometimes be understood as simply a statistical summary of a data set. With that, data assimilation is an example that extends beyond straightforward issues of description and models fill the gaps in observational data. Again there is an emphasis on the need to accurately represent soft data because the knowledge associated with the measurement comes via the model (Morrison, 2009).

Moving to the next step, simulation development, Gilbert and Troitzsch proffer that the major difference between simulation and experiment is that in the latter one is controlling the actual object of interest while in a simulation one is experimenting with a model rather than the phenomenon itself. They contend that computer simulation is similar to experimentation in that it starts with a mathematical model of the target system and application of discretizing approximations which replaces continuous variables and differential equations with values and algebraic equations (Gilbert and Troitzsch, 1999).

And then there is the question of complicated systems versus complex systems. How do they differ? These systems diverge based on the level of understanding of the system, e.g., a human system may have few parts, but it is complex because it is difficult to ascertain absolutes in the data as human systems data is organic and dynamic. Thus one cannot predict behavior the behavior of a human system with any certainty. On the other hand, a finite element model or physics-based model may be complicated due to its numerous parts, but it is not complex in that it is predictable and the data to model such a system is not soft data.

For many in the area of human behavior modeling, the concern is the basing of healthcare decisions on questionable information, soft data. Often, even the subject matter experts are apt to integrate personal opinion or judgments when providing soft data input. But this has always been the case with empirical research. Conclusions and judgments have always been proffered with the understanding that personal bias is part and parcel to the analysis.

Social scientist Simon Jackman speaks to this in his recent study on using soft data and Bayesian analysis. Jackman makes the case for allowing for the human (researcher) factor—researcher bias. He contends the research process looks like this: Prior beliefs → data → posterior beliefs in that researchers mean what they say and say what they mean, such as . . .having looked at the data, I am 95% sure that (output) is a natural product of Bayesian analysis. As such, modeling complex systems can take on more a characterization of the researcher's beliefs about a parameter in formal, probabilistic terms, rather than a statement about the repeated sampling properties of a statistical procedure. Still, this qualification regarding soft data (that it has its limitations given its fluidity and researcher bias) does not undermine its importance as it is the basis for what comprises complex systems. And as discussed earlier, human behavior modeling and human systems modeling is complex systems modeling. So then, how can human behavior and human systems modeling be enhanced?

Developing a means to measure qualitative data is a good place to start. Sokolowski and Banks have developed such a methodology (Sokolowski and Banks, Methodology 2009.) They engaged an approach to characterize social and cultural differences affecting a country's susceptibility to insurgency by placing numeric values to soft data points which simply could not be omitted from the model.

Modeling human systems is also challenging. Still, it is becoming more and more necessary in light of the need to use simulation for training healthcare providers. It has been estimated that deaths from medical errors range from 44,000 to 98,000 annually with one million injuries attributed to medical error. One can speculate as to why these numbers are so high: lack of adequate training, overworked personnel, inadequate tools, insufficient staffing. . . . With the known shortages in the healthcare community, simulation is a significant way to respond to that shortage. The Society for Simulation in Healthcare (SSH) has espoused patient safety through use of simulation-based education, collaboration, and research. Simulation training tools enable healthcare professionals to sharpen their assessment and decision-making skills without risk to patients in realistic, challenging, immersive environments that are instrumented to provide meaningful performance feedback.

CONCLUSION

Is simulation a valid and verifiable means of representing humans? Should human actions (behavior) and human physiology (systems) be reflected in computer models? This paper discussed the skepticism and reservations of professionals in various disciplines in engaging simulation in the representation of human behavior and human systems. It must be understood that M&S used in the social sciences and health sciences differ from scientific research. These disciplines do not conduct research in a physical world that is orderly and predictable in a given set of circumstances. Rather, these disciplines model complicated systems – many parts that are organized and predictable. But M&S can represent complex systems or activities. It does this by constructing models or real-world systems and simulating to observe a variety of outcomes. Significantly, this is done by the M&S professionals following a Code of Professional Ethics for Simulationists.

REFERENCES

Balci, O. Verification, Validation, and Accreditation. Proceedings of the 1998 Winter Simulation Conference, 1998. pp 41–48.

Banks CM. The History of Modeling and Simulation. In The Discipline of Modeling and Simulation: A Multidisciplinary Approach. Sokolowski, J. A., Banks, C. M., eds. Hoboken: Wiley, 2009.

Bauer JC. The Future of Medical Simulation: New Foundations for Education and Clinical Practice. White Paper for Technology Early Warning System, January 2006 *http://www.jeffbauerphd.com/TEWSMedicalSimulation.pdf* accessed 12/16/09.

Begg R, Lai DT, Palaniswami M. Computational intelligence in biomedical engineering. Boca Raton: CRC Press, 2008.

Combs D. Modeling and Simulation: Real World Examples. In Principles of Modeling and Simulation: A Multidisciplinary Approach. Sokolowski, J. A., Banks, C. M., eds. Hoboken: Wiley, 2009.

Crookall D, Zhou M. Medical and Healthcare Simulation: Symposium Overview. Simulation Gaming 2001; 32; 142 *http://sag.sagepyb.com/cgi/content/abstract/ 32/2/142* accessed 12/16/09.

Gilbert N, Troitzsch KG. Simulation for the Social Scientist. Buckingham and Philadelphia: Open University Press, 1999.

Jackman S. Bayesian Analysis for the Social Sciences. Wiley: New York. 2009.

Kyle RR, Murray WB, eds. Clinical Simulation: Operations, Engineering, and Management. Amsterdam: Elsevier, 2008.

McLeod J. But Mr. President. Proceedings of the 18th Winter Simulation Conference, 1986. pp 69–71.

McLeod J. Professional Ethics and Simulation. Proceedings of 15th Winter Simulation Conference, 1983. pp 371–373.

Morrison M. Models, Measurement, and Computer Simulation: the Changing Face of Experimentation. Philosophy Studies, (2009) 143: 33–57. DOI 10:1007/sl 1098-008-9317-y.

Oren T et al. A Code of Professional Ethics for Simulationists. Proceedings of the 2002 Summer Computer Simulation Conference, San Diego, CA, pp. 434–435.

Petty, M. Verification, Validation, and Accreditation. In Modeling and Simulation Fundamentals: Theoretical Underpinnings and Practical Domains. Sokolowski, J. A., Banks, C. M., eds. Hoboken: Wiley, 2010.

Riley RH, ed. Manual of Simulation in Healthcare. Oxford: Oxford University Press, 2008.

Sokolowski JA and Banks CM. "A Methodology to Explore the Impact of Policy Changes on Insurgencies." International Journal of System of Systems Engineering 1(3): 314–328, 2009.

Sokolowski JA, Banks CM. Investigating social dynamics and global connectivity: an agent-based modeling approach. Submitted to the Proceedings of the 2010 Winter Simulation Conference B. Johansson, S. Jain, J. Montoya-Torres, J. Hugan, and E. Yücesan, eds. (April 2010).

INDEX

Modeling and Simulation in the Medical and Health Sciences, First Edition. Edited by John A. Sokolowski and Catherine M. Banks.
© 2011 John Wiley & Sons, Inc. Published 2011 by John Wiley & Sons, Inc.